NEC4: 100 Questions and Answers

This book details some of the most important and interesting questions raised about the NEC4 family of contracts and provides clear, comprehensive answers to those questions.

Written by an NEC expert with over 20 years' experience using, advising and training others, the book has several distinctive features:

- It covers the *whole* NEC4 family
- It is written by a very experienced NEC author who explains sometimes complex issues in a simple and accessible style
- The questions and answers range from beginner level up to a masterclass level
- The questions are real life questions asked by actual NEC practitioners on real projects.

The book includes questions and answers relating to tendering, early warnings, programme issues, quality management, payment provisions, compensation events, liabilities, insurances, adjudication, termination and much more. It is essential reading for anyone working with the NEC4 family of contracts, whether professionals or students in construction, architecture, project management and engineering.

Kelvin Hughes has been a leading authority on the NEC contracts for over 20 years, during which time he has advised on numerous NEC based projects and provided over 1,700 NEC training courses at all levels from Introduction through Advanced to Masterclass. He was also Secretary of the NEC Users Group representing worldwide users of the contract, for ten years.

NEC4: 100 Questions and Answers

Kelvin Hughes

LONDON AND NEW YORK

First published 2019
by Routledge
2 Park Square, Milton Park, Abingdon, Oxon OX14 4RN

and by Routledge
52 Vanderbilt Avenue, New York, NY 10017

Routledge is an imprint of the Taylor & Francis Group, an informa business

© 2019 Kelvin Hughes

The right of Kelvin Hughes to be identified as author of this work has been asserted by him in accordance with sections 77 and 78 of the Copyright, Designs and Patents Act 1988.

All rights reserved. No part of this book may be reprinted or reproduced or utilised in any form or by any electronic, mechanical, or other means, now known or hereafter invented, including photocopying and recording, or in any information storage or retrieval system, without permission in writing from the publishers.

Trademark notice: Product or corporate names may be trademarks or registered trademarks, and are used only for identification and explanation without intent to infringe.

British Library Cataloguing-in-Publication Data
A catalogue record for this book is available from the British Library

Library of Congress Cataloging-in-Publication Data
Names: Hughes, Kelvin (Engineering consultant), author.
Title: NEC4 : 100 questions and answers / Kelvin Hughes.
    Other titles: NEC four
Description: Abingdon, Oxon ; New York, NY : Routledge, 2019. |
Identifiers: LCCN 2019000734| ISBN 9781138365247 (hbk) |
    ISBN 9781138365254 (pbk) | ISBN 9780429430855 (ebk)
Subjects: LCSH: NEC4 | Civil engineering contracts—Great Britain—
    Miscellanea. | Construction contracts—Great Britain—
    Miscellanea. | LCGFT: FAQs.
Classification: LCC KD1641 .H84 2019 | DDC 343.4107/8624—dc23
LC record available at https://lccn.loc.gov/2019000734

ISBN: 978-1-138-36524-7 (hbk)
ISBN: 978-1-138-36525-4 (pbk)
ISBN: 978-0-429-43085-5 (ebk)

Typeset in Goudy
by Swales & Willis Ltd, Exeter, Devon, UK

Contents

List of figures and tables   *xviii*
Preface   *xix*
Acknowledgements   *xxi*

Introduction to the NEC4 Contracts   1

Use of NEC4 Contracts

0.1   What are the general differences between the NEC3 and NEC4 contracts?   1

0.2   We are compiling tender documents for a project we have valued at approximately £500,000, and are wondering whether we should consider using the NEC4 Engineering and Construction Contract or the NEC4 Engineering and Construction Short Contract? What is the difference between the two, and is there any project for which the NEC4 contracts would not be applicable?   9

0.3   We wish to engage a Contractor to carry out electrical testing within existing council housing stock over a period of three years. The work is fairly straightforward. Can we use the NEC4 Term Service Short Contract or should we use the NEC4 Term Service Contract?   20

0.4   We wish to purchase large quantities of rock salt for treating roads during winter periods. Is the NEC4 Supply Short Contract sufficient, or should we use the NEC4 Supply Contract because of the large quantities?   25

0.5   When should we use the new NEC4 Design Build and Operate Contract? 27

0.6   Do the NEC4 contracts comply with the requirements of the Housing Grants, Construction and Regeneration Act 1996, and the Local Democracy, Economic Development and Construction Act 2009? 30

0.7   We wish to use the NEC4 Engineering and Construction Contract for a series of projects, and assume that the contracts include all the necessary ancillary documents to be incorporated into the contract, for example: 31

- Form of Agreement 32
- Performance bond 32
- Parent Company Guarantee 33
- Advanced Payment Bond 34
- Novation Agreement 34
- Collateral warranty 35
- Retention Bond 35

0.8   We wish to appoint a Contractor to carry out some work with the Client's design team on pre-construction value engineering and buildability options for a proposed project, for which we may eventually appoint him to be our Contractor to build the project, or we may decide to carry out a selective tendering process to select our Contractor. Can we do this with an NEC4 contract? 36

0.9   We have been using the NEC4 Engineering and Construction Contract for a number of years, and have used all the Main Options except Option D (target contract with bill of quantities), and wonder why anyone would use that option? 39

0.10   When would it be appropriate to use Option E within the NEC4 Engineering and Construction Contract? 40

0.11   We would like to use the NEC4 Engineering and Construction Contract to carry out a project using Construction Management as a procurement method. We note that Main Option F is a Management Contract, but there is no Construction Management Option? 41

0.12   We wish to use the NEC4 contracts for a number of international infrastructure projects in various locations around the world. Is that possible? *41*

Roles

0.13   What is the role of the Project Manager on an NEC4 Engineering and Construction Contract, and does he have an obligation to act impartially between the Parties? *42*

0.14   What construction discipline is best suited to being the Supervisor on an NEC4 Engineering and Construction Contract, and how much "supervising" does the Supervisor actually have to do? *48*

General

0.15   Under Clause 14.2 of the NEC4 Engineering and Construction Contract, the Project Manager or Supervisor may delegate actions. What does this mean? Is there anything that the Project Manager or Supervisor cannot, or must not, delegate? *51*

0.16   What is the meaning of the wording in Clause 10.1 and 10.2 of the NEC4 contracts, the Parties named "shall act as stated in the contract" and "act in a spirit of mutual trust and co-operation"? *52*

0.17   What is the order of precedence of documents in an NEC4 contract? *53*

0.18   What does the term "Working Areas" mean in an NEC4 Engineering and Construction Contract? Presumably this is simply, the Site? *54*

0.19   How do the NEC4 contracts deal with verbal communications? What do the words "in a form which can be read, copied and recorded" mean? *55*

0.20   We wish to name Subcontractors that the Contractor is required to use in an NEC4 Engineering and Construction Contract. How can we do that? *56*

0.21   What happens under an NEC4 Engineering and Construction Contract Option B (priced contract with bill of quantities) if *something is clearly indicated on the tender drawings, but it is missing from the Bill of Quantities?* 57

0.22 *In an NEC4 Engineering and Construction Contract Option E (cost reimbursable contract), how are insurance premiums recovered by the Contractor?* 59

0.23 *We wish to include for price fluctuations within an NEC4 Engineering and Construction Contract. How can we do that?* 59

0.24 *A new legal requirement has just come into force. The Contractor on an NEC4 Engineering and Construction Contract Option C (target contract with activity schedule) has stated that he is entitled to additional payment to recover this. Is he correct?* 61

0.25 *We wish to enter into a partnering arrangement with the Contractor within an NEC4 Engineering and Construction Contract. What can we do in order to achieve this?* 61

0.26 *Within the NEC4 Engineering and Construction Contract, when would we use Option X15 (The Contractor's Design)?* 67

0.27 *We note that the NEC4 Engineering and Construction Contract includes Option X17 (low performance damages). What is this, and how is it calculated and paid to the Client?* 68

0.28 *We do not understand NEC4 Engineering and Construction Option X18 (limitation of liability). When would this be applied?* 69

0.29 *We wish to include for Key Performance Indicators (KPIs) with an NEC4 Engineering and Construction Contract? How can we do this?* 70

0.30 *How should Z clauses be incorporated into an NEC4 contract? Are there any recommended Z clauses?* 70

0.31 *We are confused by the fact that within the NEC4 Engineering and Construction Contract, there is the Schedule of Cost Components and the Short Schedule of Cost Components? What are these for and what is the difference between the two?* 72

Contents ix

0.32   How are Subcontractors dealt with within the NEC4 Engineering
       and Construction Contract Schedules of Cost Components? 80

0.33   Some of the NEC4 Contracts include a Price List. What is this and
       how do we use it? 81

0.34   Can you explain what a Task, and a Task Order are under the
       NEC4 Term Service Contract? 83

1 Early warnings and Risk Registers   88

1.1   Is there a standard format within the NEC4 contracts for an early
       warning notice? Is there any remedy if the Project Manager or the
       Contractor fails to give an early warning? 88

1.2   What is an Early Warning Register, and what is its purpose within
       the NEC4 Engineering and Construction Contract? 92

1.3   What is the purpose of an early warning meeting within
       the NEC4 contracts? 96

2 Contractor's design: Submitting design proposals, liability
  for design, etc.   99

2.1   We wish to use the NEC4 Engineering and Construction Contract and
       the NEC4 Engineering and Construction Short Contract for a number
       of design and build projects. Can we do this and if so, how? 99

2.2   In a design and build contract using the NEC4 Engineering and
       Construction Contract, which takes precedence, the Client's
       Requirements or the Contractor's Proposals? 101

2.3   If the Project Manager on an NEC4 Engineering and Construction
       Contract accepts the Contractor's design and it is later found that the
       design will not work, or it is not approved by a third party regulator,
       who is liable? 101

2.4   We wish to novate a design consultant from the Client to the
       Contractor under an NEC4 Engineering and Construction Contract
       with full Contractor's design. Can we do this? 102

2.5 The Contractor has designed a temporary access walkway bridging across two areas of the Site. He has not submitted his design proposals for the Project Manager's acceptance before constructing the walkway, but we insist that he is in breach of contract by failing to do so. What can we do to remedy this breach? *104*

3 Time and the Accepted Programme: The submission or non-submission of a programme, float and time risk allowances, method statements, etc. **105**

3.1 Can we include liquidated or unliquidated damages in an NEC4 contract? How are these deducted in the event of delayed Completion? *105*

3.2 Do the NEC4 contracts have provision for Sectional Completion where the Client may wish to take over parts of the works before completion of the whole project? *106*

3.3 In an NEC4 Engineering and Construction Contract, what is the relationship between "Completion" and "Take over"? *106*

3.4 We have a project where we require the Contractor to provide "as built" drawings, maintenance manuals and staff training as part of the contract. How do we include this requirement within an NEC4 Engineering and Construction Contract? *107*

3.5 We have a project under the NEC4 Engineering and Construction Contract Option C (target contract with activity schedule) to build a new college where we will require the Contractor to complete wall finishings to classrooms in order that a third party directly employed by the Client will be able to install audio visual equipment fixed to the walls. How can we incorporate this requirement into the contract? *109*

3.6 The Contractor on an NEC4 Engineering and Construction Contract was required to submit a programme with his tender. He has stated that as we accepted his tender, then we have also accepted his programme, and this is therefore the first Accepted Programme. But how can it be so if it does not comply with Clause 31.2? *110*

3.7 Who owns the float in an Accepted Programme within an NEC4 Engineering and Construction Contract? *111*

3.8 The Contractor on an NEC4 Engineering and Construction Contract has failed to submit a programme to the Project Manager. What can we do to remedy this? Is there any remedy where the Project Manager fails to respond to a programme submission within the required two weeks? *113*

3.9 How often must the Contractor in an NEC4 Engineering and Construction Contract submit a revised programme? *121*

3.10 In an NEC4 Engineering and Construction Contract being carried out under Option A (priced contract with activity schedule), can we require the Contractor to bring forward Completion if the Client has a need to take over the project earlier? *123*

4 Quality management: The roles of the Parties, testing requirements, searching for Defects, etc.   125

4.1 We have identified a number of Defects on a project which a Contractor is carrying out under the NEC4 Engineering and Construction Contract, but the Contractor denies that they are Defects. How do the NEC4 contracts define a Defect? *125*

4.2 On an Engineering and Construction Contract using Option C (target contract with activity schedule), the Contractor has advised us that he is entitled to be paid for correcting a Defect. We disagree that we should be paying him for correcting Defects. *126*

4.3 What is the meaning of the term "search for a Defect" under Clause 43.1 of the NEC4 Engineering and Construction Contract? *128*

4.4 What can be done on an NEC4 Engineering and Construction Contract if the Contractor fails to correct Defects? *128*

4.5 When is the Defects Certificate issued under an NEC4 Engineering and Construction Contract, and who issues it? *128*

4.6   Our project, which we are about to carry out using the
        NEC4 Engineering and Construction Contract Option B
        (priced contract with bill of quantities), will be on completion a
        very high security building with restricted access. Once the Client has
        taken over the works, we cannot allow the Contractor to return to
        correct any Defects. How do we then deal with any Defects
        that may arise? 130

4.7   The Contractor has installed the wrong suspended ceiling tiles to the
        training rooms within our new building; however, we urgently need
        to take over the works and cannot wait for him to correct the Defect.
        How can we manage this issue? 131

5 Payment provisions: Payment and non-payment under the various options, use of Disallowed Cost, etc. 133

5.1   We wish to make use of a Project Bank Account. Can we do this
        with an NEC4 contract? 133

5.2   We note that the NEC4 Engineering and Construction Contract
        does not include provision for retention on payments to the Contractor.
        Why is this, and how can we include for retention? 134

5.3   We need to pay the Contractor on an NEC4 Engineering and
        Construction Contract, partly in US dollars and partly in the local
        currency of the country in which the project is located. How can we
        do this? 134

5.4   We wish to pay the Contractor on an NEC4 Engineering and
        Construction Contract a mobilisation payment equating to 10% of
        the value of the contract. How can we do this? 135

5.5   What is the Fee in an NEC4 Engineering and Construction
        Contract intended to cover? 135

5.6   The Contractor on an NEC4 Engineering and Construction
        Contract has failed to submit a programme to the Project Manager.
        What can we do to remedy this? 136

5.7   On an NEC4 Engineering and Construction Contract
        Option A (priced contract with activity schedule), if the Contractor

has completed an activity, should he be paid for that activity if it contains Defects? 137

5.8 We are Contractors tendering for a series of NEC4 Engineering and Construction Contracts using Option A (priced contract with activity schedule). How should we price our Preliminaries costs within the activity schedule so that we are paid correctly? 138

5.9 On an NEC4 Engineering and Construction Contract Option B (priced contract with bill of quantities), the Contractor contends that if the quantities change, he is entitled to revise his rates. Is that true? 139

5.10 What are "accounts and records" as required under an NEC4 Engineering and Construction Contract Option C, D, E and F? The Contractor has stated that there are certain accounts and records of costs that he cannot make available to the Project Manager as they are confidential, and also in some cases would breach data protection law. How can we pay the Contractor these costs? 140

5.11 What are Disallowed Costs and how are they deducted from payments due? Can Disallowed Cost be applied retrospectively to a payment made in a previous month? Is there a maximum time period for deducting Disallowed Cost? 142

5.12 How is the Contractor's Share under the NEC4 Engineering and Construction Contract Option C (target contract with activity schedule) set and calculated? 146

5.13 How are unfixed materials on or off Site dealt with under the NEC4 contracts? 148

5.14 How is interest calculated on late payments on an NEC4 Engineering and Construction Contract? 149

5.15 The Contractor on an NEC4 Engineering and Construction Contract has failed to submit an application for payment by the assessment date. The Project Manager has stated that no payment is therefore due to the Contractor this month. Is this correct? 150

xiv Contents

5.16 We have a contract under the NEC4 Engineering and Construction Contract where the Contractor has stated that he must be paid for compensation events if he has done the relevant work. Is the Project Manager required to certify payments "on account" for compensation events not yet agreed? 151

5.17 We have heard that there are some substantial changes to the payment terms within the NEC4 Professional Service Contract. What are they? 153

6 Managing compensation events: Notification, pricing and assessing compensation events, assumptions, etc. 157

6.1 How do the NEC4 contracts deal with unforeseen ground conditions on the Site? 157

6.2 How do the NEC4 contracts deal with delays and/or costs incurred by exceptionally adverse weather conditions? 159

6.3 Do the NEC4 contracts include provision for "force majeure" events? 162

6.4 How long do the Project Manager and the Contractor have on an NEC4 Engineering and Construction Contract to notify compensation events? 164

6.5 We are working on an Option A (priced contract with activity schedule) under the NEC4 Engineering and Construction Contract and have submitted quotations for compensation events to the Project Manager, but he has not responded, and he has said he will discuss them with us when we submit the Final Account. Can he do this? 166

6.6 How should a Contractor under the NEC4 Engineering and Construction Contract prepare a quotation for a compensation event? 168

6.7 We are Contractors carrying out a refurbishment contract under NEC4 Engineering and Construction Contract Option A (priced contract with activity schedule) and have had several small compensation events, each with time delays of less than one day.

However, the total cumulative effect of all these compensation events will cause a delay to the Completion Date of approximately nine days. How do we recover these delays and their associated costs under the contract? 170

6.8 We have a Consultant carrying out survey work under Option A (priced contract with activity schedule) of the NEC4 Professional Service Contract. The survey has been substantially delayed due to extremely bad weather conditions. Is such a delay the Consultant's own risk? 171

6.9 We have a compensation event in an NEC4 Engineering and Construction Contract where, in discussion with the Contractor during an early warning meeting he has stated that it is extremely difficult to price one element of the work, and we agree with him. The Contractor has suggested including a Provisional Sum within his quotation, which he can adjust later. Can he do this? 172

6.10 How should a Project Manager on an NEC4 Engineering and Construction Contract make his own assessment of a compensation event? 173

6.11 The Project Manager on an NEC4 Engineering and Construction Contract has instructed the omission of a part of the works. Can the Contractor claim for loss of profit on the omitted works? 174

6.12 We have an Engineering and Construction Contract Option B (priced contract with bill of quantities), and a description of a product in the Bill of Quantities states "or similar approved". If the Contractor then proposes a cheaper alternative (which is approved), does this saving get administered as a negative compensation event? 176

6.13 We would like to include value engineering within the NEC4 contracts and to reward the Contractor if he proposes a change that can save the Client money. Is there any provision for value engineering within the NEC4 contracts? 176

6.14 What does the term "implemented" mean within the NEC4 contracts when referring to compensation events? 179

7 Title: Title to Plant and Materials, objects and materials within the Site   180

    7.1 *During excavations for the foundations to a new building as part of a project to build a new school using the NEC4 Engineering and Construction Contract Option A (priced contract with activity schedule), the Contractor has discovered some archaeological remains. How should this discovery be dealt with under the contract?* 180

8 Indemnity, insurance and liability: Insurance requirements, claims, etc.   182

    8.1 *What are the insurance requirements within the NEC4 contracts?* 182

    8.2 *Can we include a requirement in an NEC4 contract for a Contractor or a Consultant to have Professional Indemnity (PI) insurance?* 186

9 Termination provisions: Reasons, procedures and amounts due   188

    9.1 *We are Contractors on an NEC4 Engineering and Construction Contract Option A (priced contract with activity schedule), and we wish to terminate our employment under the contract due to non-payment by the Client. Can we do this, and if so how?* 188

10 Dealing with disputes: Adjudication and tribunal   193

    10.1 *What is the difference between Option W1, W2 and W3 within the NEC4 Engineering and Construction Contract?* 193

    10.2 *When are we required to select and name the Adjudicator on our NEC4 Engineering and Construction Contract?* 201

    10.3 *What is the NEC4 Dispute Resolution Service Contract, and when should we use it?* 203

    10.4 *In an NEC4 contract, what is the tribunal?* 205

11 Preparing and assessing tenders: Completing Scope, Site
Information and Contract Data, inviting tenders, etc. 208

- *11.1 What documents do we need to compile to invite tenders for an NEC4 Engineering and Construction Contract using Option B (priced contract with bill of quantities)? 208*

- *11.2 We are preparing the Bill of Quantities for a project to be carried out under Option B (priced contract with bill of quantities). Is the preparation of the Bill of Quantities any different to any other contract? The Client wishes to make a decision about certain landscaping elements as the project proceeds, so we are intending to include Provisional Sums within the Bill of Quantities for the time being. 218*

Index 220

Figures and tables

Figures

0.1	NEC4 Engineering and Construction Contract – financial risk of Main Options	16
1.1	Suggested template for early warning notice	89
1.2	Typical Early Warning Register	94
2.1	Design and Build using the NEC4 Engineering and Construction Contract	100
4.1	Defects notification	127
4.2	Defects correction period and the Defects Certificate	129
6.1	Weather related compensation events	161

Tables

0.1	Differences between the Schedule of Cost Components and the Short Schedule of Cost Components	79
5.1	The Price for Services Provided to Date	154
5.2	The Price for Service Provided to Date	156

Preface

I have been involved with the NEC family of contracts since 1995 and in that time I have advised on numerous projects using the contracts, and over the years have carried out over 1,700 NEC based training courses. I was also Secretary of the NEC Users Group from 1996 to 2006, providing support to users of the NEC contracts including seminars and workshops, during which time I ran the Users Group Helpline answering queries on the NEC contracts from members of the Group.

During this involvement with the NEC contracts and the associated training courses, I have been asked a multitude of questions, some of which are unique, but many fall into the category of "frequently asked questions", though of course all of these questions are important to the questioner, and when asked, they all require a constructive and practical answer.

I have always felt that it would be useful to NEC practitioners if I could gather up a collection of these questions and also include the answers, but the problem is how many questions should be considered, and on that point I felt that 100 questions and answers would be a reasonable number, though I can think of many more!

Readers may be aware of one of my previous books, "NEC3 Construction Contracts: 100 Questions and Answers". This book on the NEC4 contracts does repeat some of the questions from that the previous book, though the answers of course may be slightly different, dependant on how the NEC4 contracts deal with the matter in question, but also there are many new questions, with their answers.

In dealing with the questions and answers I have primarily referred to the NEC4 Engineering and Construction Contract (ECC) as the principal reference document; many of the answers also related to the other members of the NEC4 family, for example:

- NEC4 Engineering and Construction Short Contract (ECSC)
- NEC4 Professional Service Contract (PSC)
- NEC4 Term Service Contract (TSC)
- NEC4 Term Service Short Contract (TSSC)

- NEC4 Supply Contract (SC)
- NEC4 Supply Short Contract (SSC).

I have also occasionally referred to each of the NEC4 contracts using the above abbreviations, although I tend not to use too many abbreviations, preferring to refer to the contracts by their correct title.

References to clause numbers in the book relate to the NEC4 contracts, though the clause numbering is not dissimilar to the original NEC contracts, and to subsequent NEC3 versions.

Whilst a number of practitioners still use the previous editions of the NEC contracts, particularly NEC3, the book is primarily aimed at giving guidance to NEC4 users, though the structure and contents of the contracts are again very similar, and to that end much of the advice given in this book is of use to all NEC users.

The book is intended to be of benefit to professionals who are actually using the contract, but also to students who need some awareness of the NEC contracts as part of their studies.

Readers may note the absence of case law within the text of the book. As with my previous NEC books, this is a deliberate policy on my part, for three reasons.

First, I am not a lawyer, my background being in senior commercial positions within major building contractors and more recently in senior management with a major project management company in Qatar, so I felt, and readers may concur, that I was unqualified to quote, and to attempt, a detailed commentary on any case law.

Second, there has actually been very little case law on the NEC contracts since they were first launched. This is good news in the sense that NEC has largely avoided resorting to the "tribunal" within the contracts, but lawyers will often refer to and rely on case law and precedent, and in that case I am afraid, but at the same time very pleased, that they will find very little case law with reference to the NEC contracts.

Third, and probably the most important, as a contracts consultant with significant overseas experience of all contracts including the NEC, it was always my intention that the book should attract an international readership. NEC was always conceived as an international contract, so including UK case law would probably limit it to a UK readership.

In January 2019, the NEC stated "following publication there has been discussion of, and feedback from, users and industry experts on the new features, forms and changes which were introduced within NEC4". They published a set of quite minor amendments which they recommended users should consider including in their upcoming tenders or existing contracts. As this book was already in production at that time, it does not include these amendments.

Kelvin Hughes
December 2018

Acknowledgements

As with all my previous books, I would like to extend my continued and sincere thanks to my wife Lesley, who as always gives me the love, the time, the inspiration, and the support to fulfil my life's ambitions.

Also love and thanks to my parents, Maureen and Dennis, who made me who I am, and I hope in return I continue to make them proud.

I would like to thank the publishers, the Taylor & Francis Group, for their support, and on many occasions their patience, as I completed each of my books.

Finally, to all the people who I have met over the past 44 years, who have convinced me that I made the right career choice in working within the construction industry! I hope that in return, in some small way, I have been able to add something to their careers!

Chapter 0

Introduction to the NEC4 Contracts

Question 0.1 What are the general differences between the NEC3 and NEC4 contracts?

The NEC4 contracts were launched in June 2017, and are a combination of revisions to the existing NEC3 contracts, and the introduction of two brand new members of the NEC family.

The two new NEC4 contracts are:

- NEC4 Design Build and Operate Contract (DBO)
- NEC4 Alliance Contract (ALC) (Launched in July 2017 as consultative version, final version launched June 2018).

So, the NEC4 family is now:

- NEC4 Engineering and Construction Contract (ECC)
- NEC4 Engineering and Construction Short Contract (ECSC)
- NEC4 Term Service Contract (TSC)
- NEC4 Term Service Short Contract (TSSC)
- NEC4 Design Build and Operate Contract (DBO)
- NEC4 Alliance Contract (ALC)
- NEC4 Supply Contract (SC)
- NEC4 Supply Short Contract (SSC)
- NEC4 Framework Contract (FC)
- NEC4 Engineering and Construction Subcontract (ECS)
- NEC4 Engineering and Construction Short Subcontract (ECSS)
- NEC4 Professional Service Contract (PSC)
- NEC4 Professional Service Subcontract (PSS)
- NEC4 Professional Service Short Contract (PSSC)
- NEC4 Dispute Resolution Service Contract (DRSC).

Refer to Appendix 1 – NEC4 Main and Secondary Options

More on the new contracts:

(i) NEC4 Design Build and Operate Contract (DBO).
(See also Question 0.5)

The NEC4 Design Build and Operate Contract (DBO) allows Clients to procure a fully integrated whole life delivery solution.

The benefit of a Design Build and Operate contract is that it combines responsibility for the usually separate functions – design, construction, operation and/or maintenance, all procured from a single supplier.

The Design Build and Operate Contract can include a range of different services to be provided before, during and after engineering and construction works are completed. These could include operation of the asset to achieve required performance levels, more straightforward facilities management (FM) type services, or a combination of both.

Note that the Design Build and Operate Contract does not provide for a design-build-finance-operate (DBFO) contract, i.e. the contract does not include the provisions that would be required if the supplier was funding the construction and recovering the cost during the operation phase.

(ii) NEC4 Alliance Contract (ALC)
(See also Question 0.25)

The NEC4 Alliance Contract is different from the others within the NEC4 family of contracts as it is a multi-party contract with an integrated risk and reward model.

The Alliance Contract creates a "true" alliance arrangement in which the Client and all key members of the supply chain, called "Partners", are engaged under a single contract.

The basis of the Alliance Contract is that all the Parties work together in achieving the Client objectives, and in doing so, share in the risks and benefits.

It is a multi-party contract, under which the Client and its main delivery partners, which potentially consist of Works Contractors, Consultants and Equipment Suppliers, all sign up to a single set of terms.

New Secondary Options

- Option X8 – Undertakings to the Client or Others
    This is a new Option, not formerly included within the NEC3 Engineering and Construction Contract, but in the NEC3 Professional Services Contract as "Collateral Warranty Agreements", and provides for the Contractor to give undertakings to the Client or Others as stated in the Contract Data and, if required, in the form stated in the Scope.

This may also include undertakings for example between a Subcontractor and Others if required by the Contractor. Typically, such documents are often referred to as collateral warranties.

If Parties wish to enter into collateral warranty agreements, then they need to address that matter separately, with appropriate documentation to suit the Parties.

Collateral warranties are frequently required in practice. This has remained the case even in the UK where the Contracts (Rights of Third Parties) Act 1999 apparently offers a more straightforward solution, which is also available in NEC contracts by using option Y(UK)3.

The benefit of this clause is that Clients now have the option to request the undertakings they require in the form that they require, without the need for Z clauses.

- Option X10 – Information Modelling – is a brand new Option within NEC4 and provides for Information Modelling to support the use of Building Information Modelling (BIM).

The Information Plan is submitted by the Contractor to the Project Manager who then accepts or does not accept within two weeks of receiving it.

Project Information is provided by the Contractor and used to create or change the Information Model.

The Information Model is the electronic integration of the Project Information, and other information provided by the Client and other Information Providers. The Client is liable for any fault or error in the Information Model, unless there is a Defect in the Project Information provided by the Contractor.

There is also provision within the Option for early warnings where something could affect the Information Model, and for the Contractor to include within its quotation for a compensation event where the Information Execution Plan is altered by a compensation event.

- Option X15 The Contractor's Design – which provides for Contractor's design duties to be aligned with the industry standard for designers.

Through Option X15, the Contractor's design duty has been aligned with the industry standard preferred by insurers, that is, to use the skill and care normally used by professionals designing similar works.

This Option was formerly in the NEC3 Engineering and Construction Contract, and called "Limitation of the Contractor's liability for his design to reasonable skill and care".

There are several provisions covered by this Option:

First, whilst the Scope defines what, if any design, is to be carried out by the Contractor, the core clauses within the contract are silent on the standard of care to be exercised by the Contractor when carrying out any design.

Two terms that relate to design liability are "fitness for purpose" and "reasonable skill and care".

Some contracts will limit the Contractor's liability to that of a consultant, i.e. reasonable skill and care.

The Engineering and Construction Contract does this through Option X15. If this secondary option is not chosen, the Contractor's liability for design is "fit for purpose", if it is chosen the Contractor's liability is to use "reasonable skill and care".

A retention of documents clause is provided within the Option in relation to design information to be held for the period of retention stated in the Contract Data. This is in respect of potential latent defects.

Finally, the Option deals with any requirement for professional indemnity insurance to be provided by the Contractor, using an extension to the Insurance Table.

The benefit of this Secondary Option is that it contains provisions that users and insurers frequently require, and which have otherwise been introduced to contracts by using Z clauses.

- Option X22 – Early Contractor Involvement (ECI) – Launched separately from the contracts in 2015, the ECI option is now part of the NEC4 Engineering and Construction Contract.

This is again a brand new Option, not formerly included within the NEC3 contract.

Early Contractor Involvement (ECI) is a method of appointing a Contractor at an early stage, to participate in the development of designs, budgeting and pre-construction proposals. It enables the Contractor's input to the design at a stage when significant improvements and innovation can be introduced. The same Contractor, or a different one, can then carry out and complete the construction of the project.

This procurement method helps to identify and reduce or eliminate problems and risks from a project and incorporate construction considerations at an early stage. It also provides a sound foundation for collaboration between the Parties.

The Contractor is employed in Stage One and is required to provide detailed forecasts of the total Defined Cost of the work to be done in Stage One for acceptance by the Project Manager. These forecasts are prepared on a periodic basis commencing from the starting date at intervals stated in the Contract Data. The Project Manager has one week in which to accept/not accept each forecast.

The Contractor is required, in consultation with the Project Manager, to provide forecasts of the Project Cost and submit them to the Project Manager. These forecasts are prepared on a periodic basis commencing form the starting date until completion of the whole of the works at intervals stated in the Contract Data.

The Contractor submits its design proposals for Stage Two, including a forecast of the effect of the design proposals on the Project Cost and the Accepted Programme, to the Project Manager for acceptance as stated within the submission procedure within the Scope. If the Project Manager does not accept he gives reasons and the Contractor resubmits.

The Project Manager issues a notice to proceed to Stage Two when the Contractor has obtained approvals and consents from Others, changes to the Budget have been agreed, the Project Manager and Contractor have agreed the total of the Prices for Stage Two and the Client has confirmed that the works are to proceed.

If the Project Manager does not issue a notice to proceed to Stage Two, the Client may appoint another Contractor to complete Stage Two.

If the Project Manager issues an instruction changing the Client's requirements, the Project Manager and the Contractor agree changes to the Budget within four weeks.

A budget incentive is paid to the Contractor if the final Project Cost is less than the Budget.

New features within the core clauses

- Corrupt Acts

    The NEC4 contracts now include core clauses that prohibit Corrupt Acts and provide a termination remedy in the event that a Corrupt Act is carried out.

    This has been introduced to keep the NEC4 contracts in line with Client and legislative demands, and to avoid the need for Z clause amendments.
- Contractor's proposals

    This is a new provision, where the Contractor may propose to the Project Manager that the Scope is changed to reduce the amount the Client pays to the Contractor. This could also be considered as part of a value engineering or a risk management process.

    Within four weeks of making the proposal, the Project Manager may:

    - accept the proposal and issue an instruction changing the Scope;
    - inform the Contractor that the Client is considering the proposal and instruct the Contractor to submit a quotation; or
    - inform the Contractor that the proposal is not accepted.

- Transfer of benefits (assignment)

    NEC4 contracts now contain a core clause that allows either Party to transfer the benefit or any rights under the contract to another Party.

    This has been introduced to provide for Client demands, without the need for Z clause amendments.
- Programme

    Because of the past issues of non response, or at least vague responses from Project Managers, the NEC4 Engineering and Construction Contract now has a new provision within Clause 31.3 in that, if the Project Manager fails to reply to the Contractor's programme submission accepting or not accepting it within the time allowed, the Contractor may notify the Project Manager to that effect.

If the Project Manager's failure to respond continues for a further one week after the Contractor's notification, it is treated as acceptance of the programme by the Project Manager. This will hopefully goad an "errant" Project Manager to do what the contract requires it to do!

Note also that the requirement within the NEC3 Engineering and Construction Contract for the Contractor to show in its revised programme "the effects of implemented compensation events" no longer exists in the NEC4 Engineering and Construction Contract.

This is a welcome deletion as the previous requirement caused confusion in that a compensation event is not "implemented" until the quotation has been accepted or assessed by the Project Manager.

Whilst the Completion Date is not changed until a compensation event has been implemented, there is a danger in taking the clause literally and only showing compensation events that have been implemented, so the revised programme was not reflecting reality.

- Quality management

    The NEC4 contracts introduced a requirement within the Engineering and Construction Contract for the Contractor to prepare and issue a quality management system, a quality policy statement, and a quality plan. This was already a provision within the NEC3 Professional Services Contract.

- Payment provisions

    Periodic assessments now require an application by the Contractor.

    If no application is made by the assessment date, the Contractor does not receive payment. If, however, payment is due to the Client, the Project Manager makes the assessment and certifies payment.

    This approach, which was applied in the NEC3 short forms of contract, is now applied to all NEC4 contracts.

- Schedules of Cost Components and Fee

    Some changes have been made to the Schedules of Cost Components and associated Contract Data inputs.

    The NEC4 Professional Service Contract (PSC), the NEC4 Term Service Contract (TSC) and the NEC4 Supply Contract (SC) now use Defined Cost in the same way as the Engineering and Construction Contract (ECC).

    Subcontractor costs have been moved to the Schedule of Cost Components (SCC) and payment has been made consistent across all options. The Defined Cost is the cost paid to the Subcontractor.

    The Working Areas Overhead and People Overhead in the Schedule of Cost Components and the Short Schedule of Cost Components have been removed. The relevant items are now paid as actual Defined Cost instead.

    The design overheads percentage has been removed as an addition to rates for designers.

Introduction to the NEC4 Contracts

The rules for People costs have been drafted to allow for working in different locations, not exclusively in the Working Areas. The contract now states that people whose normal place of work is in the Working Areas are to be included in Defined Cost, according to their time worked "on this contract".

The Short Schedule is used in Options A and B, and adopts a pre-priced approach for People costs in the form of People Rates to replace the cost based calculation in NEC3.

Finally, there is now only one Fee percentage in NEC4 contracts, with no separate Fee percentage for subcontracted works.

- Final assessments

    The NEC contracts have never had provision for the equivalent of a Final Account or Final Certificate which is found in other contracts, certifying that the contract has fully and finally been complied with, and that issues such as payments, compensation events and defects have all been dealt with and effectively, the contract can be closed.

    Whilst some may say that, because of the compensation event provisions there is no need for a Final Account, and the Defects Certificate confirms correction of any outstanding defects, there is no need for a Final Certificate, some questions could still remain such as when, under Option E, is it too late for the Contractor to submit a cost?

    NEC4 has introduced the new provision within its contracts where, in the case of the NEC4 Engineering and Construction Contract, the Project Manager makes a final assessment of amounts due to the Contractor, in effect giving closure, at least to financial aspects of the contract.

    The Project Manager makes an assessment of the final amount due to the Contractor and certifies a payment, no later than:

    - four weeks after the Supervisor issues the Defects Certificate or
    - thirteen weeks after the Project Manager issues a termination certificate.

    Similar to a normal payment, the Project Manager gives the Contractor details of the assessment and payment (by either Party) is made within three weeks of the assessment date or, if a different period is stated in the Contract Data, within the period stated.

- Compensation events

    NEC4 contracts now provide as a core clause the facility for additional compensation events to be added into the contract.

    This decision is made at the time the contract is prepared, with details being provided in Contract Data Part 1.

    The benefit is that Clients can now alter the standard risk profile contained in NEC4 contracts, without the need for Z clause amendments.

    A new compensation event has been added where the Project Manager notifies the Contractor that a quotation for a proposed instruction is not accepted. This is for the cost of preparing the quotation. This ensures that the

Contractor is compensated if the Project Manager requests numerous quotations for proposed instruction not all of which are accepted or instructed.

The compensation event procedure in the Short Contracts has been simplified and shortened. Rather than wait for an instruction, the Contractor submits a quotation with the compensation event notice. The Client either accepts the quotation or makes its own assessment.

- Consensual dispute resolution

NEC4 has introduced a four week period for escalation and negotiation of a dispute, which takes place prior to any formal proceedings being commenced.

This requires nominated senior representatives of each Party to meet and try to reach a negotiated solution.

It is a mandatory requirement where dispute resolution Option W1 applies, but is consensual where dispute resolution Option W2 applies (in the UK where the Housing, Grants, Construction and Regeneration Act 1996 applies).

The benefit of this is to improve the chance of a negotiated solution being reached, and maintain collaboration between the Parties.

- Dispute Avoidance Board option

The Engineering and Construction Contract now includes a dispute avoidance option W3 which can be used if the UK Housing, Grants, Construction and Regeneration Act does not apply. This is to refer any dispute to a Dispute Avoidance Board nominated by the Parties at the time the contract is formed.

The approach is that the Dispute Avoidance Board members become familiar with the project prior to any dispute arising, by making regular visits to see the project at set intervals, or when requested by the Parties.

If a dispute arises it is referred to the Dispute Avoidance Board for review. The board visit the project to discuss the issues with the Parties and help find a solution. They provide a recommendation to resolve the dispute if it is not resolved by discussion.

The dispute can only be referred to the tribunal (being either litigation or arbitration) if the recommendation provided by the Dispute Resolution Board is not accepted by the Parties.

Again, the benefit of this new option is to encourage and support the Parties in resolving any dispute or difference consensually, and to support users who wish to use this facility on their projects. The NEC3 Adjudicator's Contract is now the NEC4 Dispute Resolution Service Contract, and has been changed to allow it to be used for the appointment of board members.

- Contract Data

The Contract Data has been reformatted to make it easier to navigate and complete.

Terminology changes

The following changes in terminology have been made.

- The term "Employer", where used in most of the NEC3 contracts, has been replaced with "Client" within all the NEC4 contracts.
- The term "Scope" has been used in all contracts for the document. It describes the work being provided, replacing "Works Information", "Service Information" or "Goods Information".
- The "Risk Register" has been renamed the "Early Warning Register" to distinguish it from the Project Risk Register, which is used for wider project management purposes including health, safety and environment.
- Section 8 in NEC4 contracts has been re-titled "Liabilities and insurance" (it used to be titled "Risks and insurance"). The term "liabilities" is generally used in this section now in place of "risk" and drafting has been developed to suit.
- Secondary Option X4 has been re-titled "Ultimate holding company guarantee" instead of "Parent company guarantee", and contains provisions that recognise the existence of more complex corporate structures.
- Secondary option X12 "Partnering" has been re-titled "Multiparty collaboration" to better describe what it does.

Question 0.2 **We are compiling tender documents for a project we have valued at approximately £500,000, and are wondering whether we should consider using the NEC4 Engineering and Construction Contract or the NEC4 Engineering and Construction Short Contract? What is the difference between the two, and is there any project for which the NEC4 contracts would not be applicable?**

Many practitioners will assume that for larger projects over a certain value, the NEC4 Engineering and Construction Contract would be used, but for a lower value project the NEC4 Engineering and Construction Short Contract should be used. However, that assumption would be incorrect, as one would also need to consider the complexity of the work, the risk associated with that work, and the need for various provisions that would be in one contract but not the other.

As the notes on the NEC4 Engineering and Construction Short Contract state:

> *The NEC4 Engineering and Construction Short Contract is an alternative to Engineering and Construction Contract and is for use with contracts which do not require sophisticated management techniques, comprise straightforward work and impose only low risks on both the Client and the Contractor.*

In some ways the choice may be linked to project value in that smaller projects could be less complex, the risks could be low, and there could be a need for fewer provisions, i.e. a shorter contract.

In order to consider this question in more detail in terms of the use of the Engineering and Construction Contract or the Engineering and Construction Short Contract, and the applicability generally of NEC4 contracts, let us first briefly review the two contracts.

NEC4 Engineering and Construction Contract

The original objectives of the NEC contracts, and more specifically the Engineering and Construction Contract, were to make improvements over other contracts under three headings:

(i) Flexibility

The contract would be able to be used:

- for any engineering and/or construction work containing any or all of the traditional disciplines such as civil engineering, building, electrical and mechanical work, and also process engineering.
  Previously, contracts had been written for use by specific sectors of the industry, e.g. ICE civil engineering contracts, JCT building contracts, IChemE process contracts;
- whether the Contractor has full, some or no design responsibility.
  Previously, most contracts had provided for portions of the work to be designed by the Contractor, but with separate design and build versions if the Contractor was to design all or most of the works;
- to provide all the normal current options for types of contract such as lump sum, remeasurement, cost reimbursable, target and management contracts.
  Previously, contracts were written, primarily as either lump sum or remeasurement contracts, so there was no choice of procurement method when using a specific standard form;
- to allocate risks to suit each particular project.
  Previously, contracts were written with risks allocated by the contract drafters, and one had to be expert in contract drafting to amend the conditions to suit each specific project on which it was used;
- anywhere in the world.
  Previously, contracts included country specific procedures and legislation and therefore either could not be used or had to be amended to be used in other countries. To that end, FIDIC contracts have been seen as the only forms of contract that could be used in a wide range of countries.
  To date, the NEC contracts have been used for projects as widely diverse as airports, sports stadiums, water treatment works, housing projects, in many parts of the works, and even research projects in the Antarctic!

(ii) Clarity and simplicity

- The contract is written in ordinary language and in the present tense.
- As far as possible, NEC only uses words that are in common use so that it is easily understood, particularly where the user's first language is not English.

Previously obscure words such as "whereinbeforesaid", "hereinafter" and "aforementioned" were commonplace in contracts! The NEC contracts have few sentences that contain more than 40 words and use bullet points to subdivide longer clauses.

- The number of clauses and the amount of text are also less than in most other standard forms of contract and there is an avoidance of cross-referencing found in more traditional standard forms.
- It is also arranged in a format that allows the user to gain familiarity with its contents, and required actions are defined precisely, thereby reducing the likelihood of disputes.
- Finally, subjective words such as "fair" and "reasonable" have been used as little as possible as they can lead to ambiguity, so more objective words and statements are used.

Some critics of NEC have commented that the "simple language" is actually a disadvantage as certain clauses may lack definition and there are certain recognised words that are commonly used in contracts. There is very little case law in existence with NEC contracts, and whilst adjudications are confidential and unreported, anecdotal evidence suggests that there do not appear to be any more adjudications with NEC contracts than any other, which would tend to suggest that the criticism may be unfounded.

(iii) Stimulus to good management

This is perhaps the most important objective of the Engineering and Construction Contract in that every procedure has been designed so that its implementation should contribute to, rather than detract from, the effective management of the work. In order to be effective in this respect, contracts should motivate the Parties to proactively want to manage the outcome of the contract, not just to react to situations. NEC intends the Parties to be proactive and not reactive. It also requires the Parties and those that represent them to have the necessary experience (sadly often lacking) and to be properly trained so that they understand how the NEC works.

The philosophy is founded on two principles:

- "Foresight applied collaboratively mitigates problems and shrinks risk."
- "Clear division of function and responsibility helps accountability and motivates people to play their part."

Examples of foresight within the Engineering and Construction Contract are the early warning and compensation event procedures, the early warning provision requiring the Project Manager and the Contractor each to notify the other upon becoming aware of any matter that could have an impact on price, time or quality.

A view held by many Project Managers is that an early warning is something that the Contractor would give, and is an early notice of a "claim". This is an erroneous view as, first, early warnings should be given by either the Project Manager or the Contractor, whoever becomes aware of it first, the process being designed to allow the Project Manager and Contractor to share knowledge of a potential issue before it becomes a problem, and, second, early warnings should be notified regardless of whose fault the problem is – it is about raising and resolving the problem, not compensating the affected Party.

The compensation event procedure requires the Contractor to submit within three weeks a quotation showing the time and cost effect of the event. The Project Manager then responds to the quotation within two weeks, enabling the matter to be properly resolved close to the time of the event rather than many months or even years later.

The programme is also an important management document with the contract clearly prescribing what the Contractor must include within his programme and requiring the Project Manager to "buy into it" by formally accepting (or not accepting) the programme. The programme is then defined as the "Accepted Programme".

In total, the Engineering and Construction Contract is designed to provide a modern method for Clients, Contractors, Project Managers, Designers and others to work collaboratively and to achieve their objectives more consistently than has been possible using other traditional forms of contract. People will be motivated to play their part in collaborative management if it is in their commercial and professional interest to do so.

Uncertainty about what is to be done and the inherent risks can often lead to disputes and confrontation, but the Engineering and Construction Contract clearly allocates risks and the collaborative approach will reduce those risks for all the Parties so that uncertainty will not arise.

- Flexibility of use – the Engineering and Construction Contract is not sector specific, in terms of it being a building, civil engineering, mechanical engineering, process contract, etc. As with other contracts, it can be used for any form of engineering or construction.

    This is particularly useful where a major project such as an airport or a sports stadium can be a combination of building, civil engineering and major mechanical and electrical elements.
- Flexibility of procurement – the Engineering and Construction Contract's Main and Secondary Options, together with flexibility in terms of

Contractor design, allow it to be used for any procurement method whether the Contractor is to design all, none or part of the works. Again, other contracts do not offer this flexibility.
- Early warning – the Engineering and Construction Contract contains express provisions requiring the Contractor and the Project Manager to notify and, if required, call a "risk reduction" meeting when either becomes aware of any matter that could affect price, time or quality. Few contracts have this express requirement.
- Programme – there is a clear and objective requirement for a detailed programme with method statements and regular updates which provides an essential tool for the Parties to manage the project and to notify and manage the effect of any changes, problems, delays, etc. Whilst other contracts contain programme requirements, they do not deal with them in the same detail, and one could imply that they probably do not properly recognise the importance of a programme.
- Compensation events – this procedure is unique to NEC and requires the Contractor to price the time and "Defined Cost" effect of a change within three weeks and for the Project Manager to respond within two weeks. There is therefore a "rolling" Final Account with early settlement and no later "end of job" claims for delay and/or disruption. It is also more beneficial for the Contractor in terms of his cash flow as the Contractor is paid agreed sums rather than reduced "on account" payments, which are subject to later agreement and payment.
- Disputes – the contract encourages better relationships and there is far less tendency for disputes because of its provisions. If a dispute should arise there are clear procedures as to how to deal with it, i.e. adjudication, tribunal.

Arrangement of the Engineering and Construction Contract

The Engineering and Construction Contract includes the following sections:

Core clauses

1 General
2 The Contractor's main responsibilities
3 Time
4 Quality management
5 Payment
6 Compensation events
7 Title

8   Liabilities and insurance
9   Termination.

Main Option clauses

Option A	Priced contract with activity schedule
Option B	Priced contract with bill of quantities
Option C	Target contract with activity schedule
Option D	Target contract with bill of quantities
Option E	Cost reimbursable contract
Option F	Management contract

- Options A and B are priced contracts in which the risks of being able to carry out the work at the agreed prices are largely borne by the Contractor.
- Options C and D are target contracts in which the Client and Contractor share the financial risks in an agreed proportion.
- Options E and F are two types of cost reimbursable contract in which the financial risks of being able to carry out the work are largely borne by the Client.

Dispute resolution

- Option W1 Dispute resolution procedure (used unless the Housing Grants, Construction and Regeneration Act 1996 applies).
- Option W2 Dispute resolution procedure (used in the UK when the Housing Grants, Construction and Regeneration Act 1996 applies).
- Option W3 Dispute Resolution procedure when a Dispute Avoidance Board is the method of dispute resolution and the Housing Grants, Construction and Regeneration Act 1996 does not apply.

Secondary Option clauses

Option X1	Price adjustment for inflation
Option X2	Changes in the law
Option X3	Multiple currencies
Option X4	Ultimate holding company guarantee
Option X5	Sectional Completion
Option X6	Bonus for early Completion
Option X7	Delay damages
Option X8	Undertakings to the Client or Others
Option X9	Transfer of rights
Option X10	Information Modelling

Introduction to the NEC4 Contracts

Option X11	Termination by the Client
Option X12	Multiparty collaboration
Option X13	Performance bond
Option X14	Advanced payment to the Contractor
Option X15	The Contractor's design
Option X16	Retention
Option X17	Low performance damages
Option X18	Limitation of liability
Option X20	Key Performance Indicators
Option X21	Whole life cost
Option X22	Early Contractor Involvement
Option Y(UK)1	Project Bank Account
Option Y(UK)2	Housing Grants, Construction & Regeneration Act 1996
Option Y(UK)3	Contracts (Rights of Third Parties) Act 1999
Option Z	Additional conditions of contract.

Note: Options X19 is not used.

Note that the Local Democracy, Economic Development and Construction Act 2009 has a direct bearing on Option Y(UK)2.

The Act, specifically Part 8, amends Part II of the Housing Grants, Construction and Regeneration Act 1996, and came into force in England and Wales on 1 October 2011 and in Scotland on 1 November 2011.

It is important to note that as the Act amends the Housing Grants, Construction and Regeneration Act 1996, you have to take into account both Acts to fully understand how it applies to your contract.

Briefly, the differences between the Housing Grants, Construction and Regeneration Act 1996 and the Local Democracy, Economic Development and Construction Act 2009 include:

- A notice is to be issued by the payer to the payee within five days of the payment due date, or by the payee within not later than five days after the payment due date; the amount stated is the notified sum. The absence of a notice means that the payee's application for payment serves as the notice of payment due and the payer will be obliged to pay that amount.
- A notice can be issued at a prescribed period before the final date for payment which reduces the notified sum.
- The Contractor has the right to suspend all or part of the works, and can claim for reasonable costs in respect of costs and expenses incurred as a result of this suspension as a compensation event.
- Terms in contracts such as "the fees and expenses of the Adjudicator as well as the reasonable expenses of the other Party shall be the responsibility of the Party making the reference to the Adjudicator" will be prohibited. Much has been written about the effectiveness of such clauses, and whether they comply with the previous Act, so the new Act should provide clarity for the future.

- The Adjudicator is permitted to correct his decision so as to remove a clerical or typographical error arising by accident or omission. Previously he could not make this correction.

The Main Options

The six Main Options (Option A to Option F) enable Clients to select a procurement strategy and payment mechanism most appropriate to the project and the various risks involved (see Figure 0.1). Essentially, the main options differ in the way the Contractor is paid.

Whilst many traditional contracts are based on bills of quantities, there has been a movement away from the use of traditional bills and towards payment arrangements such as milestone payments and activity schedules, with payment based on progress achieved, rather than quantity of work done.

There is also an increasing use of target cost contracts which has been encouraged by the increasing use of partnering arrangements, the better sharing of risk and also the continued growth of NEC contracts which provide for target options. To that end, it is perhaps not surprising that in a survey carried out by the RICS (Contracts in Use Survey), Options A and C were

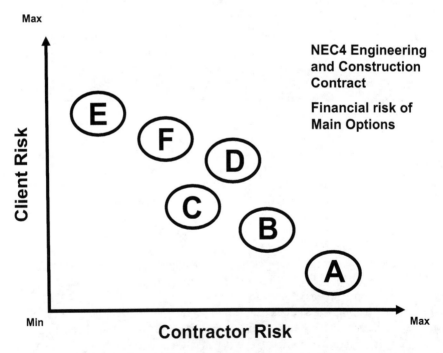

Figure 0.1 NEC4 Engineering and Construction Contract – financial risk of Main Options

found to be the most regularly used NEC4 Engineering and Construction Main Options.

Once the procurement strategy has been decided, the Main and Secondary Options can be selected to suit that strategy.

NEC4 Engineering and Construction Short Contract

As stated above, many users of the NEC4 contracts believe that the Engineering and Construction Contract is used for larger projects and the Engineering and Construction Short Contract for smaller projects.

The majority of building and engineering contracts carried out are of relatively low risk and complexity, and often but not necessarily, low value. Hence the work involved in preparing a full Engineering and Construction Contract cannot be justified in many cases.

Also, the detailed procedures and management systems in the Engineering and Construction Contract may not be necessary in many contracts. Thus, there is a definite need for a shorter contract that is simple, easy to use and more suited to the low risk and complexity, and small value type of contract.

The Latham Report Constructing the Team emphasised the need for a shorter contract, paragraph 5.20 of the report stating that *"provision should be made for a simpler and shorter minor works document"*.

Unlike other forms of contract, there is no limiting or recommended financial value for contracts under the Engineering and Construction Short Contract, because the above criteria are not in general related to a particular sum of money. It is possible to have large value, low risk and simple work carried out under the Engineering and Construction Short Contract. On the other hand it may not be appropriate to carry out complex high risk work under the Engineering and Construction Short Contract even though its monetary value is small. Several Clients have used the Engineering and Construction Short Contract very successfully for work in excess of £2m in value.

Intended users of the Engineering and Construction Short Contract range from major organisations carrying out a large number of straightforward, low risk projects, to domestic householders seeking a simple, user friendly contract to enable them to appoint a builder for a home extension. It has to be said that the great majority of users of the contract fall into the former category.

Since the Engineering and Construction Short Contract has been drafted for use on straightforward and low risk work, many matters covered by the Engineering and Construction Contract have been omitted. Whilst the intention has been to cater for anything that might go wrong on the simpler project, it is always possible that something may occur that is not covered in the Engineering and Construction Short Contract. Such matters will then have to be dealt with by the common law of contract. A comparison with the Engineering and Construction Contract is given below.

Differences between the NEC4 Engineering and Construction Contract and the NEC4 Engineering and Construction Short Contract

Detailed comparison with the clauses of the NEC4 Engineering and Construction Contract is outside the scope of this book.

However there are some important differences of a general nature.

For example, the Project Manager and Supervisor are absent, but the Client has powers to delegate as necessary to suit his organisation and his particular purposes.

Also there are no Main Options – instead the flexibility and allocation of risk is covered within the Price List.

There are no Secondary Options, the reason for this is that these would probably not be used in the type of contract for which the Engineering and Construction Short Contract is designed. However, retention and delay damages, both secondary options in the Engineering and Construction Contract, have been included in the clauses of the Engineering and Construction Short Contract.

The number of definitions has been reduced, and since there are no main options, the Prices and the Price for Work Done to Date have standard definitions with the flexibility being inherent in the use of the Price List.

Some clauses that are unlikely to be used have been omitted but if required can be incorporated by appropriate drafting of the Scope or the use of additional conditions of contract. For instance, no mention is made of health and safety requirements, but any specific requirements can be included in the Scope.

There is a requirement that the Client is to accept the Contractor's design before he can start work on the work designed although detailed procedures for acceptance have not been included. Similarly, there is no procedure for acceptance of Subcontractors but responsibility of the Contractor for the performance of Subcontractors is clearly stated.

In the Engineering and Construction Contract there is a prescribed list of items that must be included in the Contractor's programme. All of these may not be necessary for a project using the Engineering and Construction Short Contract, and therefore any detailed requirements for the programme within the Engineering and Construction Short Contract are to be stated in the Scope.

The payment procedure in the Engineering and Construction Short Contract is different from that in the Engineering and Construction Contract. In the Engineering and Construction Short Contract the Contractor is required to submit an application for payment after which the Client is obliged to pay after making any corrections he considers necessary. There is no payment certificate as in the Engineering and Construction Contract.

Also, the definition of Contractor's Cost is much simplified. There is no Schedule of Cost Components; instead, cost is defined in terms of payments made by the Contractor for four different items.

- people employed by the Contractor
- Plant and Materials
- work subcontracted by the Contractor and
- Equipment.

The list of compensation events has been reduced. The weather event has been simplified in terms of time lost.

Assessment of compensation events is similar to that of the Engineering and Construction Contract except that effectively the rates in the Price List are used where possible.

The section on Title is much abbreviated and the title to surplus materials on the site is now with the Contractor rather than the Client.

Section 8 on insurance etc. has been completely restructured to include the equivalent of force majeure events.

Reasons which give rights to termination in Section 9 have been reduced and some of the detailed procedures on termination have been omitted.

Contract strategy

The Engineering and Construction Short Contract does not give the Client the choice of contract strategy in the form that is provided in the Engineering and Construction Contract.

However, the Price List can be used to produce a lump sum contract or a bill of quantities based contract. Target contracts and cost reimbursable contracts are not provided for as they are regarded as unsuitable and too complex for this type of work. Similarly, Management Contracts will not be used on the type of work which the Engineering and Construction Short Contract is designed for.

When the Client prepares the contract he must decide the strategy he wishes to employ and hence who should carry the risks of quantities, pricing etc.

In response to the last part of the question, the author is not aware of any project for which the NEC contracts would not be applicable.

Some contracts were specifically written and published for specific types of project; for example, the IChemE contracts are for use on process contracts and have specific Schedules for pre-installation and performance tests guarantees and procedures, and spare parts for future maintenance which are required for such contracts.

The IChemE contracts also have variants for Lump Sum (Red Book), Cost Reimbursable (Green Book) and Target Cost (Burgundy Book) contracts, so in that case they have some of the flexibility of the NEC contracts.

One could get into a prolonged debate about which contract is best, but the author believes that the NEC contracts are probably still better contracts – one just has to make sure all the requirements and constraints for the works, or the service, are properly provided for within the Scope.

Question 0.3   We wish to engage a Contractor to carry out electrical testing within existing council housing stock over a period of three years. The work is fairly straightforward. Can we use the NEC4 Term Service Short Contract or should we use the NEC4 Term Service Contract?

Again, as stated previously with other NEC4 contracts, whether one uses the NEC4 Term Service Short Contract or the NEC4 Term Service Contract is not dictated by the size of the project, but by the complexity of the work, the risk associated with that work, and the need for various provisions that would be in one contract but not in the other.

In order to consider this question, let us firstly briefly review the two contracts.

NEC4 Term Service Contract

The Term Service Contract is designed for use in a wide variety of situations where a Contractor is appointed to carry out, as its name suggests, a defined service over a predetermined period of time, referred to in the contract as the "service period" starting at the "starting date".

When deciding whether to use the Term Service Contract one must recognise that this is not a contract to provide a project. The principle is that it is based on providing a service, i.e. maintaining an existing condition for a period of time (term) to permit the Client's continuing use of a facility.

It does not normally include the improvement of an existing condition of an asset – that would comprise a project. However, a modest amount of improving the condition of an asset, sometimes called "betterment", may sometimes be sensibly included in a Term Service Contract.

Another consideration is that there is no equivalent of the "Site" or "Working Areas" as in the Engineering and Construction Contract. The contract refers to the "Affected Property".

The "Scope" describes the service to be provided by the Contractor. It also includes full details of where and how it is to be provided and any constraints placed upon the Contractor.

Examples of the use of the Term Service Contract include:

- maintenance of highways in a particular area
- periodic inspection and reporting on structures, e.g. bridges and tunnels
- cleaning of streets in an urban area
- refuse collection and disposal
- maintaining public parks and landscape areas
- maintaining mechanical and electrical installations in buildings
- maintenance of water courses such as rivers and canals
- the provision of security personnel for an installation, site or building.

Examples of other and more complex applications may be:

- servicing and maintaining airport and sea terminal buildings
- maintaining lifts in a group of hospitals
- the provision of data processing services by a computer systems company for a number of years
- carrying out long-term maintenance as part of facilities management contracts.

The form and structure of the NEC4 Term Service Contract is similar to other NEC4 contracts, so users will be immediately familiar with early warning, compensation events, etc.

The Client is represented by a "Service Manager" who administers the contract on behalf of the Client in the same way that the Project Manager and Supervisor do in the Engineering and Construction Contract.

Other differences between the Term Service Contract and the Engineering and Construction Contract include:

The Contractor's plan

The Term Service Contract replaces the NEC4 Engineering and Construction Contract's programme with the Contractor's plan. The reason for this is that in a service contract much more emphasis is placed on how the Contractor proposes to provide the service throughout the service period, rather than the timing of his activities to provide the Works, as in the NEC4 Engineering and Construction Contract.

Affected Property

The Site and Working Areas of the Engineering and Construction Contract are replaced by the Affected Property which is a defined term in the core clauses.

Core clauses

1. General
2. The Contractor's main responsibilities
3. Time
4. Quality management
5. Payment
6. Compensation events
7. Use of equipment, Plant and Materials
8. Liabilities and insurance
9. Termination.

Main Option clauses

There are three Main Options, one of which must be selected. They are:

Option A   Priced contract with price list
Option C   Target contract with price list
Option E   Cost reimbursable contract.

Each of these Main Options has a different allocation of risk between Contractor and Client.

Option A has the greatest financial risk for the Contractor in that he is bound by the rates and prices in the Price List regardless of the actual cost to him of providing the services. The Price List is a combination of lump sum items and quantity related items. Payments to the Contractor (defined as the Price for Service Provided to Date) consist of lump sums for those items that have been completed and amounts for the other items calculated as quantities of work completed multiplied by the rates, i.e. by admeasurement.

Option C is a cost reimbursable contract in which the Contractor is paid his Defined Cost together with the "Contractor's share".

The latter is calculated by comparing the Defined Cost at various stages in the provision of the services, with a target price (the Prices) which is calculated from the rates and prices in the Price List. The difference between the Prices and the Price for Services Provided to Date is divided into share ranges. The Contractor's share is calculated for each of the share ranges and the total of these is the total Contractor's share payable to the Contractor.

Certain "Disallowed Cost" defined in the core clauses is deducted to arrive at the Defined Cost which determines the payments to be made to the Contractor in Options C and E. The Fee is added to Defined Cost – this is intended to broadly cover the Contractor's head office overheads and profit. The target (the Prices) is adjusted to allow for compensation events as they are assessed in accordance with the contract.

Option E is a cost reimbursement contract similar to Option C but without the incentive to the Contractor in the form of the Contractor's share. But there is again provision for Disallowed Cost. It is therefore intended for use only where the risks are high and where the service to be provided by the Contractor cannot be defined with any certainty at the start of the contract. It would also be suitable where the service required is of an experimental nature. Pricing in the Price List for Option E contracts is used only for estimating and budgeting purposes.

Dispute resolution

- Option W1
- Option W2.

The Price List

- The Price List must be prepared for each contract. A pro-forma Price List is included in the Guidance Notes.

It will be seen from the above that the Price List consists of two kinds of entries:

- Lump sum items and
- Remeasurable items.

The Price List may be compiled by either the Client or the Contractor, with the exception of the pricing. Pricing should be completed by the Contractor or negotiated with him.

Where the Contractor is paid, say, an amount each month for the service he is providing, the description of the work covered by the sum of money is entered in the "Description" column, and the rate for each month is entered in the "Rate" column. The total number of months is entered in the "Expected Quantity" column. Thus the flexibility in payment methods inherent in the main Options of the Engineering and Construction Contract is provided in the Term Service Contract by means of the different ways in which the Price List can be used.

Secondary Options

Option X1	Price adjustment for inflation
Option X2	Changes in the law
Option X3	Multiple currencies
Option X4	Ultimate holding company guarantee
Option X8	Undertakings to the Client or Others
Option X10	Information modelling
Option X11	Termination by the Client
Option X12	Multiparty collaboration
Option X13	Performance bond
Option X17	Low service damages
Option X18	Limitation of liability
Option X19	Termination by either Party
Option X20	Key Performance Indicators
Option X21	Whole life cost
Option X23	Extending the Service Period
Option X24	The accounting periods
Option Y(UK)1	Project Bank Account
Option Y(UK)2	The Housing Grants, Construction & Regeneration Act 1996
Option Y(UK)3	Contracts (Rights of Third Parties) Act 1999
Option Z	Additional conditions of contract.

Term Service Short Contract

The NEC4 Term Service Short Contract should again be used for the appointment of a supplier for a period of time to manage and provide a service. This contract is an alternative to the NEC4 Term Service Contract and is for use with contracts that do not require sophisticated management techniques, comprise straightforward work and impose only low risks on both the Client and the Contractor.

Complete package

The Term Service Short Contract is published as a complete package including:

- pre-printed forms for
    - Contract Data
    - Contractor's Offer and Client's Acceptance
    - Price List
    - Scope
    - Conditions of Contract.

When all the forms have been completed for a particular contract the package will comprise the complete contract document, together with the drawings and anything referred to in the Scope.

The most important notes on the use of the document are included in grey bordered boxes.

The Parties

The main parties referred to in the Term Service Short Contract are the:

- Client
- Client's Agent (if appointed)
- Contractor
- Adjudicator.

The Client and the Contractor are the Parties to the contract.

The Client/Client's Agent

Unlike the Term Service Contract, the Term Service Short Contract does not include the role of Service Manager, all actions being between the Client (or his delegated agent) and the Contractor

If the Client appoints an agent it can be either from his own staff, or from an outside body. His role is to manage the contract for the Client to achieve the Client's objectives for the completed service. He is normally appointed at the feasibility stage of the service, his duties then including advising on procurement, cost planning and programme matters.

If the Client has set limits upon his level of authority, for instance agreeing the value of compensation events, he must ensure that there is an efficient and speedy authorisation procedure to allow the agent to exceed these limits.

The Adjudicator

The Adjudicator is appointed jointly by the Client and the Contractor for the contract. The name of the Adjudicator is inserted into the Contract Data. If the Contractor does not agree with the choice a suitable alternative should be appointed before the Contract Date.

The Adjudicator becomes involved only when either contracting Party refers a dispute to him. As an independent person he is required to give a decision on the dispute within stated time limits. If either Party does not accept his decision they may proceed to the tribunal (litigation or arbitration). Payment of the Adjudicator's fees is shared by the Parties.

Question 0.4   We wish to purchase large quantities of rock salt for treating roads during winter periods. Is the NEC4 Supply Short Contract sufficient, or should we use the NEC4 Supply Contract because of the large quantities?

The NEC4 Supply Contract is used for the procurement, supply and delivery of high value goods and associated services, which can range through items such as transformers, turbines, trains, process plant, etc., together with related services that may be required as part of the contract, such as design and also specific delivery requirements.

"Procurement, supply and delivery" means that these items are obtained and delivered to a Delivery Place on a Delivery Date, but there is no fixing/installation element to the contract.

What is classed as "high value" is not based on a finite figure, it is really the decision of the Client, but there is a need for a comprehensive contract that will cover all aspects of the procurement and supply process for those goods and services.

The Parties within the NEC4 Supply Contract are the Purchaser and the Supplier, with a Supply Manager appointed and named in the Contract to represent the Purchaser.

The structure is very similar to the other NEC4 contracts, so there is provision for early warnings, programmes submitted for acceptance, and also compensation events to manage changes.

Core clauses

1. General
2. The Supplier's main responsibilities
3. Time
4. Quality management
5. Payment
6. Compensation events
7. Title
8. Liabilities and insurance
9. Termination, resolving and avoiding disputes.

There are no Main Options, but the following Secondary Options:

Option X1	Price adjustment for inflation
Option X2	Changes in the law
Option X3	Multiple currencies
Option X4	Ultimate holding company guarantee
Option X7	Delay damages
Option X10	Information modelling
Option X11	Termination by the Purchaser
Option X12	Multiparty collaboration (not used with Option X20)
Option X13	Performance bond
Option X14	Advanced payment to the Supplier
Option X17	Low performance damages
Option X18	Limitation of liability
Option X20	Key Performance Indicators (not used with Option X12)
Option X21	Whole life cost
Option X25	Supplier warranties
Option Y(UK)1	Project Bank Account
Option Y(UK)3	The Contracts (Rights of Third Parties) Act 1999
Option Z	Additional conditions of contract.

The NEC4 Supply Short Contract (not as often quoted, the "Short Supply Contract") should be used for procurement of goods probably under a single order or on a batch order basis, and is for use with contracts that do not require sophisticated management techniques and impose only low risks on both the Purchaser and Supplier.

The principle of the complexity of the requirements and whether the risks are high or low is always the deciding factor as to whether to use an NEC4 contract or its "Short Contract" version, rather than simply the value of the contract, as is so often the case with other contract families. It is not about value, it is about what you are trying to do and what factors need to be considered in doing it and achieving those objectives.

The titles of the core clauses within the Supply Short Contract are the same as the Supply Contract, and for that matter most of the other NEC4 contracts:

1 General
2 The Supplier's main responsibilities
3 Time
4 Quality management
5 Payment
6 Compensation events
7 Title
8 Liabilities and insurance
9 Termination, resolving and avoiding disputes.

The NEC4 Supply Short Contract is structured very similarly to the other NEC4 Short Contracts, i.e.:

- Contract Data
- The Supplier's Offer and the Purchaser's Acceptance
- Price List
- Scope.

Again, as with the other Short Contracts there is shortened provision for early warnings, programmes submitted for acceptance, and also compensation events to manage changes.

For the purchase of large quantities of rock salt, it is not the quantity of material that would need to be considered in deciding whether to use the NEC4 Supply Contract or the Supply Short Contract, but the nature of how the deliveries are made, and any other factors such as risk, etc. It is suggested that the NEC4 Supply Short Contract is probably the appropriate choice.

Question 0.5 When should we use the new NEC4 Design Build and Operate Contract?

In general terms a Design Build and Operate (DBO) contract is a project delivery model in which a single Contractor is appointed to design and build a project, and then to operate it for a period of time, often referred to as the "concession period".

It is often used for highways projects, especially where a toll road or bridge is incorporated, and also for water treatment and irrigation schemes, where the Contractor may also, at least initially, finance the project.

The common form of a Design Build and Operate contract is a Public Private Partnership (PPP), in which a Public Client (e.g. government or public agency) enters into a contract with a Private Contractor to design, build and then operate the project, while the Client finances the project and retains ownership.

This differs from a Design Build Finance and Operate (DBFO) contract in which the Contractor also finances the project and leases it back to the Client for an agreed period (perhaps 30 years) after which the development reverts to the Client.

It also differs from the traditional design and build contract in that it includes operation and maintenance of the completed works, which means that the Contractor's duties and responsibilities to the Client do not end at completion of construction, but continue through a defined and usually prolonged operational term.

In theory, this encourages the Contractor to develop a project with its long-term performance in mind from the outset, rather than just considering the efficiency of its construction, as the Contractor will be responsible for any high operating, maintenance or repairs bills.

However, it ties both the Client and the Contractor into a very long term relationship that can be difficult to price. As a result, Contractors may price considerable risk into their tenders, and so the Client may not always achieve a best value outcome.

The NEC4 Design Build and Operate Contract (DBO) is a new addition to the NEC family and allows Clients to procure a fully integrated whole life delivery solution. It reflects the increasing demand for contracts extending into the operational phase.

Note that the Design Build and Operate Contract does not provide for a design-build-finance-operate (DBFO) contract, i.e. the contract does not include the provisions that would be required if the supplier was funding the construction and recovering that cost during the operation phase.

New and existing assets

The new DBO contract is not only intended for the "traditional" DBO approach, where the client requires a new facility or asset to be designed, constructed and then operated by a Contractor. It is also available for situations where a client wants to have an existing facility or asset operated by the contractor while it is being upgraded or extended.

In such situations, the timing of the design and construction phase would normally be stated by the client, but may be determined by the contractor – which will undertake the work at a point to suit the performance requirements of the overall service.

The new DBO contract provides for option A (priced contract with price list), option C (target contract with price list) and option E (reimbursable contract).

All the standard NEC provisions for communications, early warning and events are retained along with some other features added into the NEC4 suite.

The management of the work through plans and programmes for the construction and operation phases are still important.

Introduction to the NEC4 Contracts

Incentives and productivity

A key issue for the Client is identifying and incentivising performance and productivity of the total asset. This may be through any or all of the traditional NEC approaches, including the reduction of payments for poor performance.

The new DBO contract brings incentives and productivity options together into one schedule and requires the client to identify and develop the approaches it wishes to use.

Structure of the contract

Core clauses

1. General
2. The Contractor's main responsibilities
3. Time
4. Quality management
5. Payment
6. Compensation events
7. Use of equipment, Plant and Materials
8. Liabilities and insurance
9. Termination.

Resolving and avoiding disputes

1. Option W1
2. Option W2.

There are no Main Options, but the following Secondary Options:

Option X3	Multiple currencies
Option X4	Ultimate holding company guarantee
Option X8	Undertakings to the Client or Others
Option X9	Transfer of rights
Option X10	Information modelling
Option X13	Performance bond
Option X14	Advanced payment to the Contractor
Option X18	Limitation of liability
Option X23	Extending the Service Period
Option Y(UK)2	The Housing Grants, Construction and Regeneration Act 1996
Option Y(UK)3	The Contracts (Rights of Third Parties) Act 1999
Option Z	Additional conditions of contract.

Schedule of cost components

Contract data

Part 1 – Data provided by the Client

Part 2 – Data provided by the Contractor.

Question 0.6   Do the NEC4 contracts comply with the requirements of the Housing Grants, Construction and Regeneration Act 1996, and the Local Democracy, Economic Development and Construction Act 2009?

If one just examines the Core Clauses of the NEC4 contracts, the simple answer is no, the NEC4 contracts do not comply with the requirements of the Housing Grants, Construction and Regeneration Act 1996 and Local Democracy, Economic Development and Construction Act 2009, and the simple reason is because the NEC4 contracts are for use anywhere in the world, therefore they do not include any country specific clauses.

However, Main Option Y(UK)2: The Housing Grants, Construction and Regeneration Act 1996 is applicable to UK contracts where the Housing Grants, Construction and Regeneration Act 1996 applies.

Note that Option Y(UK)2 only deals only with the payment aspects of the Act – adjudication under the Act is covered by Option W2.

Whilst this book avoids discussing NEC4 specifically in connection with UK law, it is worth mentioning the Local Democracy, Economic Development and Construction Act 2009 which has a direct bearing on Option Y(UK)2. The Local Democracy, Economic Development and Construction Act 2009, specifically Part 8, amends Part II of the Housing Grants, Construction and Regeneration Act 1996, and came into force in England and Wales on 1 October 2011 and in Scotland on 1 November 2011.

It is important to note that as the Local Democracy, Economic Development and Construction Act amends the Housing Grants, Construction and Regeneration Act 1996, you have to take into account both Acts to fully understand how it applies to your contract.

The Act applies to all construction contracts including:

- construction, alteration, repair, maintenance, etc.
- all normal building and civil engineering work, including elements such as temporary works, scaffolding, site clearance, painting and decorating
- consultants' agreements concerning construction operations
- labour only contracts
- contracts of any value.

It excludes:

- extraction of oil, gas or minerals
- supply and fix of plant in process industries, e.g. nuclear processing, power generation, water or effluent treatment
- contracts with residential occupiers
- header agreements in connection with PFI contracts
- finance agreements.

The Housing Grants, Construction and Regeneration Act 1996 only applied to contracts in writing, but the Local Democracy, Economic Development and Construction Act 2009 now also applies to oral contracts.

Briefly, the differences between the Housing Grants, Construction and Regeneration Act 1996 and the Local Democracy, Economic Development and Construction Act 2009 include that within the latter Act:

- A notice is to be issued by the payer to the payee within five days of the payment due date, or by the payee within not later than five days after the payment due date, the amount stated is the notified sum. The absence of a notice means that the payee's application for payment serves as the notice of payment due and the payer will be obliged to pay that amount.
- A notice can be issued at a prescribed period before the final date for payment which reduces the notified sum.
- The Contractor has the right to suspend all or part of the works, and can claim for reasonable costs in respect of costs and expenses incurred as a result of this suspension as a compensation event.
- Terms in contracts such as "the fees and expenses of the Adjudicator as well as the reasonable expenses of the other Party shall be the responsibility of the Party making the reference to the Adjudicator" will be prohibited. Much has been written about the effectiveness of such clauses, and whether they comply with the previous Act, so the new Act should provide clarity for the future.
- The Adjudicator is permitted to correct his decision so as to remove a clerical or typographical error arising by accident or omission. Previously he could not make this correction.

Question 0.7 We wish to use the NEC4 Engineering and Construction Contract for a series of projects, and assume that the contracts include all the necessary ancillary documents to be incorporated into the contract, for example:

- Form of Agreement
- Performance bond

- Parent Company Guarantee
- Advanced Payment Bond
- Novation Agreement
- Collateral warranty
- Retention Bond

The simple answer to this question is that the NEC4 contracts do not have the ancillary documents themselves which would need to be incorporated into an NEC4 contract.

Whilst many NEC4 practitioners may allege that this is a clear and major oversight on the part of the NEC4 drafters, one must always remember that the NEC documents when first conceived were always intended to be for use on projects worldwide and in that sense it is impossible to include *every* conceivable ancillary document for *every* contract, compliant with the law in *every* country.

Let us consider each of the above in turn, using the NEC4 Engineering and Construction as a reference document:

Form of Agreement

There is no standard Form of Agreement for completion and signature within the NEC4 contracts, as the drafters of the contracts have always stated that Parties enter into contracts in many different ways, by executing Forms of Agreement, by exchange of letters, etc.

Whilst this is probably true, it would have assisted the Parties if there had been a standard Form of Agreement that they could use if they so wish, though within the Engineering and Construction Contract Guidance Notes there is a sample form.

The difficulty with not having a Form of Agreement embedded within the Contract is that the Client has to write his own, which requires them to summarise the documents that comprise the contract and also to consider whether the contract is to be executed as a deed. In that respect they must be very cautious that all the documents that are intended to form the contract are included in the Form of Agreement.

Having said this, there is however a signature page within the Contract Data within the various Short Contracts within the NEC4 family, for example the Engineering and Construction Short Contract has a page that consists of the Contractor's Offer and the Client's Acceptance.

Performance bond

A performance bond is an arrangement whereby the performance of a contracting Party (the Principal) is backed by a third party (the Surety), which could be a bank, insurance company or other financial institution, that should the Principal fail in his obligations under the contract, normally due to the Principal's insolvency, the Surety will pay a pre-agreed sum of money to the other contracting Party (the Beneficiary).

Secondary Option X13 provides for the Contractor to give a performance bond to the Client, provided by a bank or insurer. The bank or insurer that provides the performance bond must be accepted by the Project Manager.

If a performance bond is required from the Contractor it should be provided by the Contract Date. The amount of the bond, usually 10% of the contract value, must be stated in Contract Data Part 1 and the form of the bond must be in the form stated in the Scope.

Normally, the value of the performance bond does not reduce, but they should have an expiry date, which could be completion of the project, the Defects Date or may even include the six, ten or twelve year limitation period following completion of the works to cover any liability for potential latent defects, the expiry date being defined within the Scope.

Parent Company Guarantee

This form of guarantee is given by a parent company (or ultimate holding company) to guarantee the proper performance of a contract by one of its subsidiaries (the Contractor), who whilst having limited financial resources himself, may be owned by a larger, more financially sound parent company.

In most cases, it is the parent company that provides the guarantee, but sometimes, particularly when the parent company is in another country, a parent company may just be a company further up the chain within the group, perhaps the national parent who has sufficient assets to provide the required guarantee.

Secondary Option X4 provides for a guarantee provided by an ultimate holding company, which is normally used as an alternative to a performance bond (Secondary Option X13).

Such a guarantee is cheaper than a performance bond, as the Contractor will normally just charge an administration fee rather than the case of performance bond, where the Contractor is actually paying a premium for an insurance policy; but it may give less certainty of redress because it is not supplied by an independent third party so it is dependent on the survival, and the ability to pay, of the ultimate holding company.

An ultimate holding company is normally defined as an entity:

(a) of which the Contractor (or any member if the Contractor is an Association of Persons) is a branch, subsidiary or other similar entity; or
(b) which directly or indirectly exercises management control over the Contractor (or any member if the Contractor is an Association of Persons).

Advanced Payment Bond

Advanced payments are appropriate when the Contractor will incur significant "up front" costs before it starts receiving payments, for example in pre-ordering specialist materials, plant or equipment.

Secondary Option X14 provides for this payment, and if an advanced payment bond is required it is issued by a bank or insurer that the Project Manager has accepted, the bond being in the amount that the Contractor has not repaid.

Advance payment bonds are a helpful security when an advance payment is made to a Contractor for works to be performed. The Project Manager must accept the provider of that bond.

The amount of the repayment instalments is stated in Contract Data Part 1.

Novation Agreement

While the principle of Design and Build agreements is that the Client prepares his requirements and sends them to the tendering Contractors and they prepare their proposals to match the Client's requirements, in reality, in nearly half of Design and Build contracts the Client has already appointed a design team that prepares feasibility proposals and initial design proposals before the tenders are invited.

Outline planning permission and sometimes detailed permission may have also been obtained for the scheme before the Contractor is appointed.

Each Contractor then tenders on the basis that the Client's design team will be novated or transferred to the successful tendering Contractor who will then be responsible for appointing the team and completing the design under a new agreement.

This process is often referred to as "novation", which means "replace" or "substitute" and is a mechanism where one Party transfers all its obligations and benefits under a contract to a third party. The third party effectively replaces the original Party as a Party to that contract, so the Contractor is in the same position as if he had been the Client from the commencement of the original contract.

This approach allows the Client and his advisers time to develop their thoughts and requirements, consider planning consent issues, then when the design is fairly well advanced, the Designers can be passed to the successful Design and Build Contractor.

The NEC4 contracts do not include pro forma novation agreements, but, if used, it is critical that the wording of these agreements be carefully considered, as there are many badly drafted agreements in existence. Novation can be by a signed agreement or by deed.

As with most contracts, there should be consideration, which is usually assumed to be the discharge of the original contract and the original Parties' contractual obligations to each other. If the consideration is unclear, or where there is none, the novation agreement should be executed as a deed.

In practice, the best way is to have an agreement drafted between the Client and the Designer covering the pre-novation period, and a totally separate agreement drafted between the Contractor and the Designer for the post-novation period.

When a contract is novated, the other (original) contracting party must be left in the same position as he was in prior to the novation being made. Essentially, a novation requires the agreement of all three Parties.

As this book is intended as a guide for NEC4 practitioners worldwide, it is not intended to examine specific legal cases within the UK, as they may not apply on an international basis, but there is certainly case law available in terms of design errors in the pre-novation phase being carried through as the liability of the Contractor in the post-novation stage, and also the accuracy and validity of site investigations and the Client's and the Contractor's obligations in terms of checking and taking ownership of such information.

Collateral warranty

A collateral warranty is an agreement associated with another contract, collateral meaning "additional but subordinate" or "running side by side", the collateral warranty being entered into by a party to the primary contract.

Collateral warranties are different from traditional contracts, in that the two Parties to the warranty do not have any direct commercial relationship with each other. They are useful for creating ties between third parties, but the main disadvantage of collateral warranties is the expense of providing them, particularly where there are many interested third parties such as with housing and commercial properties. When one also considers that Parties can change their names or business status during the life of a warranty, then even a medium-sized project can involve 30 or more warranties.

There is a new Option X8, not formerly included within the NEC3 Engineering and Construction Contract, but in the NEC3 Professional Services Contract as "Collateral Warranty Agreements", and provides for the Contractor to give undertakings to Others as stated in the Contract Data and, if required, in the form stated in the Scope.

This may also include undertakings between a Subcontractor and Others if required by the Contractor. Typically such documents are often referred to as collateral warranties.

The Client prepares the undertakings and sends them to the Contractor for signature, the Contractor signs or arranges for the Subcontractor to sign them, and returns them to the Client within three weeks.

Retention Bond

On particularly large projects it may be practical to release retention monies early with a bond in place to protect the Parties in the event of defects arising.

The Client may select Option X16 and retain a proportion of the Price for Work Done to Date once it has reached any retention free amount, the retention percentage and any retention free amount being stated in Contract Data Part 1.

Following Completion of the whole of the works, or the date the Client takes over the whole of the works whichever happens earlier, the retention percentage is halved, and then the final release is upon the issue of the Defects Certificate.

It is important to note that retention is held against undiscovered defects and not incomplete work.

A Retention Bond is provided by a bank, and guarantees that if contract retention sums are paid to the Contractor early and the Contractor is in breach of its obligations, and subject to a notice from the Client confirming that breach and making a demand, the bank will pay a predetermined sum to the Client without the Client having to prove entitlement to the predetermined amount.

If stated in the Contract Data, or if agreed by the Client, the Contractor provides a Retention Bond provided by a bank or insurer, accepted by the Project Manager, and in the form stated in the Scope.

Question 0.8   We wish to appoint a Contractor to carry out some work with the Client's design team on pre-construction value engineering and buildability options for a proposed project, for which we may eventually appoint him to be our Contractor to build the project, or we may decide to carry out a selective tendering process to select our Contractor. Can we do this with an NEC4 contract?

There are two alternatives to provide for this arrangement.

The first alternative is to employ the Contractor for the initial pre-construction period by appointing him as a Consultant using the NEC4 Professional Service Contract or the NEC4 Professional Service Short Contract, then use the NEC4 Engineering and Construction Contract for the construction works.

The Contractor is initially providing a service under the NEC4 Professional Service Contract or the NEC4 Professional Service Short Contract, which can be defined within the Scope, which specifies and describes the service and states any constraints on how the Contractor who will actually construct the project can then later be appointed direct by negotiation or by an open or selective tendering process.

Although this sounds a very simple process, procurement is rarely that simple, and obviously, in suggesting this option the Client must consider, dependent on location, the relevant procurement law that may apply and will restrict how Consultants, Contractors and others are selected and appointed, particularly for public funded projects.

Introduction to the NEC4 Contracts 37

Many believe that the NEC4 Professional Service Contract is just for appointing Consultants, but as the contract itself clearly states "this contract is for the appointment of a supplier to provide professional services"; that "supplier" can be a Consultant, a Contractor, or even a Supplier in the normal sense of the word of goods and materials, particularly if they are specialist such as suppliers of electrical switchgear, process plant, specialist glazing, etc.

The structure of the Professional Service Contract is very similar to the other NEC4 contracts, so there is provision for choice of main and secondary options, early warnings, programmes submitted for acceptance, and also compensation events to manage changes.

Core clauses

1. General
2. The Consultant's main responsibilities
3. Time
4. Quality management
5. Payment
6. Compensation events
7. Rights to material
8. Liabilities and insurance
9. Termination.

Main Option clauses

Option A	Priced contract with activity schedule
Option C	Target contract
Option E	Cost reimbursable contract.

Resolving and Avoiding Disputes

Option W1
Option W2.

Secondary Option clauses

Option X1	Price adjustment for inflation
Option X2	Changes in the law
Option X3	Multiple currencies
Option X4	Ultimate holding company guarantee
Option X5	Sectional Completion
Option X6	Bonus for early Completion
Option X7	Delay damages
Option X8	Undertakings to Others

38 Introduction to the NEC4 Contracts

Option X9	Transfer of rights
Option X10	Information Modelling
Option X11	Termination by the Client
Option X12	Multiparty collaboration
Option X13	Performance Bond
Option X18	Limitation of Liability
Option X20	Key Performance Indicators
Option Y(UK)1	Project Bank Account
Option Y(UK)2	Housing Grants, Construction & Regeneration Act 1996
Option Y(UK)3	The Contracts (Rights of Third Parties) Act 1999
Option Z	Additional conditions of contract.

The problem with this alternative is that, even if you remain with the one Contractor throughout, <u>two</u> contracts are required, the NEC4 Professional Service Contract or the NEC4 Professional Service Short Contract, then the NEC4 Engineering and Construction Contract.

The second alternative is to use the new Option X22: Early Contractor Involvement within the NEC4 Engineering and Construction Contract. For this option the whole process takes place via <u>one</u> contract, the NEC4 Engineering and Construction Contract.

This is a new Option, not formerly included within the NEC3 contract.

Option X22 commences with four identified and defined terms:

1. Budget – the items and amounts stated in the Contract Data
2. Project Cost – the total paid by the Client for the items stated in the Budget
3. Stages One and Two – as stated in the Scope
4. Pricing Information – the information which specifies how the Contractor prepares its assessment of the Prices for Stage Two.

The Contractor is required to provide detailed forecasts of the total Defined Cost of the work to be done in Stage One for acceptance by the Project Manager. These forecasts are prepared on a periodic basis commencing from the starting date at intervals stated in the Contract Data.

The Project Manager has one week in which to accept/not accept each forecast.

The Contractor is required, in consultation with the Project Manager, to provide forecasts of the Project Cost and submit them to the Project Manager. These forecasts are prepared on a periodic basis commencing from the starting date until completion of the whole of the works at intervals stated in the Contract Data.

The Contractor submits its design proposals for Stage Two, including a forecast of the effect of the design proposals on the Project Cost and the Accepted Programme, to the Project Manager for acceptance as stated within the submission procedure within the Scope. If the Project Manager does not accept he gives reasons and the Contractor resubmits.

Introduction to the NEC4 Contracts   39

The Project Manager issues a notice to proceed to Stage Two when the Contractor has obtained approvals and consents from Others, changes to the Budget have been agreed, the Project Manager and Contractor have agreed the total of the Prices for Stage Two and the Client has confirmed that the works are to proceed.

If the Project Manager does not issue a notice to proceed to Stage Two, the Client may appoint another Contractor to complete Stage Two.

If the Project Manager issues an instruction changing the Client's requirements, the Project Manager and the Contractor agree changes to the Budget within four weeks.

A budget incentive is paid to the Contractor if the final Project Cost is less than the Budget.

Question 0.9   We have been using the NEC4 Engineering and Construction Contract for a number of years, and have used all the Main Options except Option D (target contract with bill of quantities), and wonder why anyone would use that option?

Main Option D (target contract with bill of quantities) is probably the least used of the six Main Options.

As with Option B (priced contract with bill of quantities), the Bills of Quantities are prepared by the Client.

This option is normally used where the Client knows what he wants and is able to define it clearly through the Scope, and measure it within the Bills of Quantities, but there are likely to be changes in the quantities, that may or may not be considered as compensation events.

The Contractor tenders a price based on the Bills of Quantities. This price, when accepted, is then referred to as the "target". The original target is referred to as the Total of the Prices at the Contract Date.

The assessment of payments is again the same as for an Option C contract:

- The target price includes the Contractor's estimate of Defined Cost plus other costs, overheads and profit to be covered by his Fee.
- The Contractor tenders his Fee in terms of percentages to be applied to Defined Cost.
- During the course of the contract, the Contractor is paid Defined Cost plus the Fee.
- The target is adjusted for compensation events and also for inflation (if Option X1 is used).
- On Completion, the Contractor is paid (or pays) his share of the difference between the final total of the Prices and the final Payment for Work Done to Date according to a formula stated in the Contract Data. If the final Payment for Work Done to Date is greater than the final total of the Prices, the Contractor pays his share of the difference.

- As with Option B the target is generated as a remeasurement, based on the Bill of Quantities, though changes which are compensation events are inserted into the Bill of Quantities as lump sums rather than on a remeasurement basis.
- The target is adjusted for compensation events and also for inflation (if Option X1 is used).

The concern with Option D amongst Clients is that as the purpose of the Bills of Quantities is essentially just as a tender document to create the target, they spend time and money producing a Bill of Quantities yet, once the Contractor is appointed and the target is established, the Bill of Quantities plays only a secondary part as it is not used for payments.

Question 0.10   When would it be appropriate to use Option E within the NEC4 Engineering and Construction Contract?

Option E is a cost reimbursable contract with the Contractor being reimbursed Defined Cost plus the Fee.
  It should be used:
- where the scope of work is uncertain, e.g. some refurbishment projects
- where extreme flexibility is required, e.g. for enabling work
- where a high level of Client involvement is envisaged
- for emergency work
- where trials or work of an experimental nature are carried out.

The option allows development of the design as the works proceed and permits maximum flexibility in allocation of design responsibility.
  A cost reimbursable contract should be used where the definition of the work to be done is inadequate even as a basis for a target price and yet an early start is required. In such circumstances the Contractor cannot be expected to take risks. It carries minimum risk and is reimbursed its Defined Cost plus Fee, subject only to a number of constraints designed to motivate efficient working.
  A criticism of cost reimbursable contracts such as Option E is that it gives the Contractor very little incentive to reduce costs; however, a cost reimbursable contract should be used where the definition of the work to be done is inadequate for the Contractor to price and yet an early start is required. In such circumstances the Contractor cannot be expected to take financial risks. It carries minimum risk and is reimbursed its Defined Cost plus Fee, subject to any deduction for Disallowed Costs.
  Under Option E of the Engineering and Construction Contract, the Contractor is reimbursed its Defined Cost plus a Fee, basically covering its offsite overheads and profit. This Fee is calculated by applying the fee percentage, given at tender by the Contractor in Contract Data Part 2, to appropriate Defined Cost.

Although Option E is a cost reimbursable contract, there is an obligation on the Contractor to provide a regular forecast of the Total of the Prices which advises the Client of its potential outturn cost.

Another criticism of Option E is that the tendering Contractors do not actually price the works; they just price their fee percentage together with any appropriate Equipment rates, Working Areas overheads percentage and manufacture and fabrication rates and overheads percentages. In addition tenderers may be required to notionally agree to a cost plan. This makes it difficult to assess cost reimbursable tenders, but again the basis of cost reimbursable contracts is that the Client bears most of the risk.

The Contractor is required to submit forecasts at the intervals stated in the Contract Data of the Total of the Prices from the starting date to the Completion of the whole of the works.

The Contractor is also required to advise the Project Manager on the practical implications of the design of the works and on subcontracting arrangements.

Question 0.11 We would like to use the NEC4 Engineering and Construction Contract to carry out a project using Construction Management as a procurement method. We note that Main Option F is a Management Contract, but there is no Construction Management Option?

The principle with the Construction Management procurement method is that the Client has *direct* contracts with the various "Contractors", otherwise known in other contracts as "Works Contractors" or "Trade Contractors", and he has a party such as a Client's Agent or Representative to act on his behalf in dealing with programme, instructions, payments, changes, etc.

It is correct that there is no Construction Management Option, each of the Engineering and Construction Contract Main Options assumes that the Parties to the contract are a single Client and a single Contractor to build a project.

However, it is possible to create a Construction Management arrangement by using the NEC4 Engineering and Construction Contract as the contract between the Client and each Contractor, the collective assembly of all the Contractors being appointed to build the single project.

The Client's Agent or Representative can then be employed as a Consultant under the NEC4 Professional Service Contract and be named as the Project Manager within each contract between the Client and the Contractor.

Question 0.12 We wish to use the NEC4 contracts for a number of international infrastructure projects in various locations around the world. Is that possible?

The NEC contracts were always conceived and created as being for use on *any* type of project (buildings, civil engineering, process engineering, etc.) in *any* location throughout the world, hence the use of non "country specific" drafting throughout the contracts, which is extended into all the members of the NEC4 family.

An example of "country specific" drafting is the JCT contracts and their specific relevance to UK based projects, and the law of England and Wales (they have to be amended for use outside these countries) including within their provisions for UK health and safety (CDM), fair contracts (Housing Grants, Contraction and Regeneration Act 1996 and the Local Democracy, Economic Development and Construction Act 2009) and other legislation.

With the NEC4 contracts, Clients in any country can include Secondary Options to provide for any country specific legislation.

So the answer is, yes, the NEC4 contracts can be used for a wide range of projects in any location throughout the world and in fact have been used for projects as widely diverse as airports, sports stadiums, power stations, water treatment works, housing projects, in many parts of the world, and even research projects in the Antarctic, with great success!

Question 0.13 What is the role of the Project Manager on an NEC4 Engineering and Construction Contract, and does he have an obligation to act impartially between the Parties?

The Project Manager is appointed by the Client and manages the contract on his behalf.

Other forms of contract tend to name an Engineer or Architect, who acts for the Client, and in addition to having design responsibilities also administers the contract. The NEC4 Engineering and Construction Contract separates the role of design from that of managing the project and administering the contract.

The Project Manager is named in Contract Data Part 1 and may be appointed from the Client's own staff or may be an external consultant. The Project Manager is a named individual, not a company.

It is vital that the Client appoints a Project Manager who has the necessary knowledge, skills and experience to carry out the role, which includes acceptance of designs and programmes, certifying payments and dealing with compensation events.

With regard to impartiality, the Project Manager's role is to act on behalf of the Client in managing the project, and the contract for him.

In that sense he is not truly impartial, but for example when called to accept or not accept something such as the Contractor's design, his proposed Subcontractor or his programme, the Project Manager either accepts or gives his reasons for not accepting, so he has a duty to comply with the contract and to act fairly and "in a spirit of mutual trust and co-operation", but in doing so to act on behalf of the Client.

If the Project Manager is seen not to be complying with the contract, and the Contractor is dissatisfied, he obviously has the right to refer any matter to adjudication.

Disciplines that have carried out the Project Manager role to date include Engineers from a civil engineering, structural or process background, Architects, Building Surveyors and Quantity Surveyors, in addition to Project Managers themselves.

It is also vital that the Client gives the Project Manager full authority to act for him, particularly when considering the time scales imposed by the contract. If the Project Manager has to seek approvals and consents from the Client, then he must do so, and the appropriate authority must be given, in compliance with the contractual time scales.

If the approval process is likely to be lengthy, it is critical that both the Client and his Project Manager set up an accelerated process to comply with the contract or that the time scales in the contract are amended. The former is the vastly preferred method in order that NEC4 will work to its full effect. Although the Project Manager manages the contract at the post-contract stage, he is usually appointed pre-contract to deal with matters such as feasibility issues, advising on design, procurement, cost planning tendering and programme matters.

Note that, unlike many other contracts, the Engineering and Construction Contract does not name the Quantity Surveyor, so this would need to be considered in dealing with financial matters pre- and post-contract.

Either the Project Manager may carry out the role of the Quantity Surveyor himself or he may delegate to another.

The role of a Project Manager within the Engineering and Construction Contract is different to the usual concept of a Project Manager and it is important that this difference is recognised as it is a traditional area of misunderstanding amongst users of the contract.

There should only be one Project Manager for the project, though the Project Manager can delegate responsibilities to others after notifying the Contractor, and may subsequently cancel that delegation (Clause 14.2).

As with any other notification under the contract, if the Project Manager (or the Supervisor) wishes to delegate, the notice to the Contractor must be in a form that can be read, copied and recorded, i.e. by letter or email, not by verbal communication such as a telephone call. The delegation may be due to the Project Manager being absent for a period due to holidays or illness, or because he wishes to appoint someone to assist him in his duties.

Although the contract is not specific, the notice should identify who the Project Manager is delegating to, what their authority is, and how long the delegation will last, so the Contractor is in no doubt as to who has authority under the contract. The authority is shared, and the Project Manager may still take an action which they have delegated.

Although the Project Manager manages the contract at the post-contract stage, he is usually appointed pre-contract to deal with matters such as feasibility issues, advising on design, procurement, cost planning tendering and programme matters. Note that, unlike many other contracts, the Engineering and Construction Contract does not name the Quantity Surveyor, so this would need to be considered in dealing with financial matters pre- and post-contract. Either the Project Manager may carry out the role of the Quantity Surveyor himself or delegate to another.

The role of a Project Manager within the Engineering and Construction Contract is different to the usual concept of a Project Manager and it is important that this difference is recognised as it is a traditional area of misunderstanding amongst users of the contract.

If the Client wishes to replace the Project Manager or the Supervisor he must first notify the Contractor (Clause 14.4) including naming his replacement.

Project Manager duties

The Project Manager:

General

10.1	acts as stated in the contract
10.2	acts in a spirit of mutual trust and co-operation
13.3–13.5	replies to a communication within the period for reply
13.6	issues its certificates to the Client and Contractor
13.8	may withhold acceptance of a submission by the Contractor
14.2	may delegate any of its actions
14.3	may give an instruction which changes the Scope or a Key Date
15.1	gives an early warning
15.2	prepares a first Early Warning Register
15.2	instructs the Contractor to attend a first early warning meeting
15.2	may instruct the Contractor to attend an early warning meeting
15.4	revises the Risk Register at each early warning meeting
16.2	accepts or does not accept the Contractor's proposal
17.1	notifies the Contractor as soon as aware of an ambiguity or inconsistency
17.1	notifies the Contractor as soon as aware of any illegal or impossible requirement
19.1	gives an instruction to the Contractor stating how it is to deal with the event.

Contractor's main responsibilities

21.2	accepts or does not accept the Contractor's design
23.1	accepts or does not accept the Contractor's design of Equipment

24.1	accepts or does not accept the Contractor's replacement person
24.2	may instruct the Contractor to remove an employee
25.3	assesses the additional cost incurred by the Client for Contractor failing to meet a Key Date
26.2	accepts or does not accept the Contractor's proposed Subcontractor
26.3	accepts or does not accept the proposed subcontract documents.

Time

30.2	decides the date of Completion
31.1	receives the Contractor's first programme
31.3	accepts or does not accept the Contractor's programme
32.2	accepts or does not accept the Contractor's revised programme
34.1	may instruct the Contractor to stop or not to start any work
35.3	certifies the date upon which the Client takes over a part of the works
36.1	instructs the Contractor to submit a quotation for acceleration

Testing and Defects

40.2	accepts or does not accept the Contractor's quality policy statement and quality plan
41.6	assesses the cost of repeating a test after a Defect is found
44.4	arranges for the Client to allow the Contractor access to any part of the works
45.1	may change the Scope to accept a Defect
45.2	accepts or does not accept the Contractor's quotation for accepting a Defect
46.1	assesses the cost of having a defect corrected by other people
46.2	assesses the cost to the Contractor of correcting the Defect.

Payment

50.1	assesses the amount due at each assessment date
50.2	considers any application for payment from the Contractor
50.6	corrects any wrongly assessed amount due
51.1	certifies a payment within one week of each assessment date
53.1	makes an assessment of the final amount due

Compensation events

60.1(17)	notifies a correction to an assumption
61.1	notifies the Contractor of a compensation event
61.3	is notified by the Contractor of a compensation event
61.4	notifies the Contractor of its decision that the Prices, the Completion Date and the Key Dates are not to be changed

46 Introduction to the NEC4 Contracts

61.5	notifies the Contractor that it did not give an early warning
61.6	states assumptions about an event too uncertain to be forecast reasonably
62.1	may instruct the Contractor to submit alternative quotations
62.3	replies within two weeks of receipt of a Contractor's quotation
62.4	instructs the Contractor to submit a revised quotation
62.5	extends the time allowed for submitting or replying to a quotation
62.6	replies to the Contractor's notification
63.2	may agree with the Contractor rates and lump sums to assess the change in Prices
63.7	assesses a compensation event as if the Contractor had given early warning
63.11	corrects the description of the Condition for a Key Date
64.1	assesses a compensation event
64.2	assesses the programme for remaining work
64.3	notifies the Contractor of its assessment of a compensation event
64.4	replies to the Contractor's notification
65.1	may instruct the Contractor to submit a quotation for a proposed instruction
65.1	implements each compensation event by notifying the Contractor.

Title

70.2	permits the Contractor to remove Plant and Materials
72.1	allows the Contractor to leave Equipment in the works
73.1	instructs the Contractor in respect of objects of value or historical interest.

Liabilities and insurance

84.1	receives the Contractor's insurance policies and certificates
86.1	submits the Client's insurance policies and certificates to the Contractor.

Termination

90.1	is notified of a reason for termination and issues termination certificate
92.2	informs the Contractor when the Client no longer requires Equipment on the Site.

Main Option clauses

Option A

55.3	receives the Contractor's revision of the activity schedule
63.16	may agree a new rate with the Contractor.

Introduction to the NEC4 Contracts 47

Option B

| 60.6 | corrects mistakes in the bill of quantities |
| 63.16 | may agree a new rate with the Contractor. |

Option C

11.2(26)	decides Disallowed Cost
20.3	receives advice from the Contractor on practical implications of design and subcontracting arrangements
20.4	prepares with the Contractor forecasts of total Defined Cost for the works
26.4	accepts or does not accept the Contractor's proposed pricing information
50.9	reviews the records made available by the Contractor
52.3	is allowed to inspect the Contractor's accounts and records
54.1	assesses the Contractor's share
54.3	makes a preliminary assessment of the Contractor's share at Completion
54.4	makes a final assessment of the Contractor's share
93.4	assesses the Contractor's share after certifying termination.

Option D

11.2(26)	decides Disallowed Cost
20.3	receives advice from the Contractor on practical implications of design and subcontracting arrangements
20.4	prepares with the Contractor forecasts of total Defined Cost for the works
26.4	accepts or does not accept the Contractor's proposed pricing information
52.3	is allowed to inspect the Contractor's accounts and records
54.5	assesses the Contractor's share
54.7	makes a preliminary assessment of the Contractor's share at Completion
54.8	makes a final assessment of the Contractor's share
60.6	gives an instruction to correct a mistake in the Bill of Quantities
93.5	assesses the Contractor's share after certifying termination.

Option E

11.2(26)	decides Disallowed Cost
20.3	receives advice from the Contractor on practical implications of design and subcontracting arrangements
20.4	prepares with the Contractor forecasts of total Defined Cost for the works

26.4	accepts or does not accept the Contractor's proposed pricing information
52.4	is allowed to inspect the Contractor's accounts and records.

Option F

11.2(27)	decides Disallowed Cost
20.3	receives advice from the Contractor on practical implications of design and subcontracting arrangements
20.4	prepares with the Contractor forecasts of total Defined Cost for the works
26.4	accepts or does not accept the Contractor's proposed pricing information
52.4	is allowed to inspect the Contractor's accounts and records.

Secondary Options

X4	accepts or does not accept a proposed alternative guarantor
X10	gives an early warning, accepts or does not accept a first Information Execution Plan
X13	accepts or does not accept a performance bond
X14	accepts or does not accept a bank or insurer
X16	accepts or does not accept a bank or insurer
X20	receives Contractor's report against KPIs
X20	receives Contractor's proposals for improving performance
X21	receives Contractor's proposal that the Scope is changed and consults with the Contractor
X22	accepts or does not accept the Contractor's forecasts of total Defined Cost, issues or does not issue a notice to proceed to Stage Two, discusses with the Contractor ways of dealing with the change to the Budget, makes a preliminary and final assessment of the budget incentive
Y(UK)1	accepts or does not accept details of the banking arrangements, receives the Authorisation.

Question 0.14 What construction discipline is best suited to being the Supervisor on an NEC4 Engineering and Construction Contract, and how much "supervising" does the Supervisor actually have to do?

In order to properly consider this question, we need first to consider the role and responsibilities of the Supervisor.

Introduction to the NEC4 Contracts   49

The Supervisor is appointed by the Client, their role being to manage issues regarding quality, testing and defects, on the Client's behalf.

The Supervisor also has another role to play in marking Plant and Materials which are outside the Working Areas, the Contractor prepares them for marking as the Scope requires, and the Supervisor then marks it as for this contract, title then passing to the Client.

As with the Project Manager, the Supervisor is named in Contract Data Part 1 and may be appointed from the Client's own staff or may be an external consultant, and again, as with the Project Manager, the Supervisor is a named individual not a company.

It is possible that, provided he has sufficient expertise and time available, the Project Manager and the Supervisor could be the same person.

It is also not unusual for the Client's Designer and the Supervisor to be the same person as he is well qualified, having prepared the design, to inspect the works to make sure the Contractor has complied with it.

The title Supervisor is not normally found in other contracts, many seeing the role as similar to that of a Clerk of Works or Resident Engineer, and in many ways the role is similar, but it is important to recognise that those roles are usually delegated by others within their respective contracts, therefore giving the Clerk of Works or Resident Engineer little or no ultimate responsibility.

The normal role of a Supervisor is to take charge of, and direct people or activities, though in effect he does not actually supervise in that sense as Clause 44.1, for example, requires that the Contractor "corrects a Defect whether or not the Supervisor has notified it" clearly requiring the Contractor to be proactive in dealing with Defects and not rely on the Supervisor to tell him what is defective and what to do about it, which would be the traditional role of a supervisor in a company.

Disciplines that have carried out the Supervisor role to date depend on the type of project, but they include Engineers from a civil engineering, structural or process background for highways and infrastructure projects, Architects and other design disciplines for building projects, Building Surveyors for building refurbishment projects, and also Clerks of Works. It really depends on the type of work what discipline the Supervisor comes from.

It is important to recognise that the Supervisor does not act on behalf of the Project Manager, he represents and reports directly to the Client and has his own responsibilities and obligations within the contract, which may be summarised as follows:

- The Supervisor carries out and/or witnesses tests being carried out by the Contractor or some other party.
- The Supervisor may instruct the Contractor to search for a Defect. This is common to other contracts where it is referred to as "opening up" or "uncovering", the principle being if no Defect is found the matter is dealt with as a compensation event.

- The Supervisor notifies the Contractor of each Defect as soon as he finds it.
- The Supervisor issues the Defects Certificate at the later of the defects date and the end of the last defect correction period. This is a very important action as once the Defects Certificate has been issued, the Contractor is no longer required to provide the insurances under the contract.

    Also, dependent on which secondary options have been selected, any retention (Clause X16) is released back to the Contractor, if a Defect included in the Defects Certificate shows low performance, the Contractor pays low performance damages (Clause X17), and the Contractor no longer has to report performance against KPIs (Clause X20).

    The Defects Certificate is issued on the later of the defects date and the last defect correction period, not when all defects have been corrected. In this case, the Defects Certificate may show Defects that the Contractor has not corrected.
- The Supervisor marks Plant and Materials as for the contract if the contract identifies them for payment. Marking Plant and Materials can involve making a physical mark on them to denote that the Supervisor has seen them, but can include compiling an inventory, photographic records, etc.

    The Contractor is required to prepare the Plant and Materials for marking as required by the Scope, which can include setting them aside from other stock, protection, insurances and any vesting requirements. Once they have been marked, any title the Contractor has to them passes to the Client.

    Again, as with the Project Manager, there should only be one Supervisor, who may then delegate duties and authorities to others. On a large project it is common for various personnel to be responsible for checking mechanical, electrical, structural, finishings, landscape elements, etc., but again there should be one Supervisor who delegates duties to others in respect of these elements. Each then reports back to the single Supervisor.

Supervisor duties

The Supervisor:

10.1	acts as stated in the contract
10.2	acts in a spirit of mutual trust and co-operation
13.3–13.5	replies to a communication within the period for reply
13.6	issues certificates to the Project Manager, the Client and the Contractor
14.2	may delegate any of its actions
41.3	notifies the Contractor of tests and inspections
41.5	does its tests without causing unnecessary delay

42.1	notifies the Contractor that Plant and Materials have passed tests
43.1	may instruct the Contractor to search for a Defect
43.2	notifies the Contractor of each Defect which it finds
44.3	issues the Defects Certificate
71.1	marks Equipment, Plant and Materials outside the Working Areas.

In answer to the last part of the question, the Supervisor is not there to supervise the Contractor, but to supervise all matters regarding quality, testing and defects on behalf of the Client. The word Supervisor is a term used in the past within civil engineering contracts, i.e. Supervising Officer.

Question 0.15 Under Clause 14.2 of the NEC4 Engineering and Construction Contract, the Project Manager or Supervisor may delegate actions. What does this mean? Is there anything that the Project Manager or Supervisor cannot, or must not, delegate?

As stated above, there should only be one Project Manager and one Supervisor for each project, though they can both delegate responsibilities to others after notifying the Contractor, and may subsequently cancel that delegation (Clause 14.2), any reference to an action of either the Project Manager or the Supervisor including an action by his delegate.

As with any other notification under the contract, if the Project Manager (or the Supervisor) wishes to delegate under Clause 14.2, the notice to the Contractor must be in a form that can be "read, copied and recorded", i.e. by letter or email, not by verbal communication such as a telephone call. The delegation may be due to the Project Manager temporarily being absent for a period due to holidays or illness, or because he wishes to appoint someone to assist him in his duties, possibility for the duration of the contract.

Although the contract is not specific, the notice should identify who the Project Manager is delegating to, what their authority is, and how long the delegation will last, so the Contractor (and the Client) is in no doubt as to who has authority under the contract.

It is important to recognise that in delegating the Project Manager is *sharing* an authority with the delegate who will then represent him – he is not *passing responsibility* on to them. The authority is shared, but ultimately total responsibility will remain with the Project Manager as the principal.

It is also important to note that one can only delegate outwards from the principal, i.e. no one can assume authority, possibly because, within an organisation, they are senior to the Project Manager. If the Client wishes to replace the Project Manager or the Supervisor he must first notify the Contractor of the name of the replacement before doing so.

Question 0.16 What is the meaning of the wording in Clause 10.1 and 10.2 of the NEC4 contracts, the Parties named "shall act as stated in the contract" and "act in a spirit of mutual trust and co-operation"?

The first clause of all the NEC contracts (Clause 10.1) has always required the Parties, and their agents, e.g. Project Manager, Supervisor, Service Manager, to act as stated in the contract, and in a spirit of mutual trust and co-operation.

This mirrors Sir Michael Latham in his report "Constructing the Team" when he recommended that the most effective form of contract should include "a specific duty for all Parties to deal fairly with each other and in an atmosphere of mutual co-operation".

This has been slightly changed within the NEC4 contracts in that the former Clause 10.1 has now been split into two separate clauses, Clauses 10.1 and 10.2 in the ECC states:

- Clause 10.1: *The Parties, the Project Manager and the Supervisor shall act as stated in the contract.*
- Clause 10.2: *The Parties, the Project Manager and the Supervisor act in a spirit of mutual trust and co-operation.*

If we examine the clauses, first and rather curiously, Clause 10.1 is written in the future tense, which is unusual for NEC contracts in that they are written in the present tense.

Notwithstanding that, is it necessary to state that the Parties, *the Project Manager and the Supervisor* are required to act as stated in the contract?

What difference does it make if that clause was absent or deleted? Would they *not* have to act as stated in the contract?

Second, what does it mean that the Parties, the Project Manager and the Supervisor are required to act in a spirit of mutual trust and co-operation?

This second requirement mirrors Sir Michael Latham in his report "Constructing the Team" when he recommended that the most effective forms of contract should include "*a specific duty for all parties to deal fairly with each other and in an atmosphere of mutual co-operation*".

This clause has often been viewed with some confusion, and for those who have spent many years in the construction industry, with a degree of scepticism.

Most practitioners state that their understanding of the clause is that the Parties should be non-adversarial toward each other, acting in a collaborative way and working for each other rather than against each other, and in reality that is what the clause requires, which is essentially correct.

However the difficulty is, that if a party does *not* act in a spirit of mutual trust and co-operation what can another party do? The answer is that the clause is almost unenforceable as it is virtually impossible to define and quantify the breach or the ensuing damages that flow from the breach.

In addition, the Contractor is not contractually related to the Project Manager or Supervisor, so either would be unable to take action directly against the other for breach of contract other than through the Client.

To that end, as Clause 10.1 is a fairly redundant clause, and Clause 10.2 is unenforceable, Clients have been seen to insert a Z clause to delete the requirements; however, it is recommended that it should remain in the contract, if merely viewed as a "statement of good intent".

In effect, a clause requiring Parties to act in a certain spirit will probably not, on its own, have any real effect. Within the Engineering and Construction Contract it is the clauses that follow within the contract that require early warnings, clearly detailed programmes which are submitted for acceptance, and a structured change management process, that actually create and develop that level of mutual trust and co-operation rather than simply inserting a statement within the contract requiring the Parties to do so.

Question 0.17   What is the order of precedence of documents in an NEC4 contract?

The NEC4 contracts, unlike most other construction contracts, do not provide a priority or hierarchy of documents within the contract clauses.

However, upon through examination of the contract, one can establish that there is a hierarchy between various documents.

For example, if the Contractor has design responsibility and there is an inconsistency between the Scope provided by the Client and the Scope provided by the Contractor, which takes precedence?

The answer can be found in Clauses 11.2(6) and 60.1(1).

Defects are defined in Clause 11.2(6) as:

- a part of the works which is not in accordance with the Scope; or
- a part of the works designed by the Contractor which is not in accordance with the applicable law or the Contractor's design which the Project Manager has accepted.

Clause 60.1(1): The Project Manager gives an instruction changing the Scope except

- a change made in order to accept a Defect; or
- a change to the Scope provided by the Contractor for his design which is made at
    - the Contractor's request or
    - in order to comply with other Scope provided by the Client.

So, one can see that Scope provided by the Client prevails over the Scope, or any changes to the Scope provided by the Contractor.

Acceptance of design

The Contractor is required to submit the particulars of his design as the Scope requires to the Project Manager for acceptance. The particulars of the design in terms of drawings, specification and other details must clearly be sufficient for the Project Manager to make the decision as to whether the particulars comply with the Scope and also, if relevant, the applicable law.

The Scope may stipulate whether the design may be submitted in parts, and also how long the Project Manager requires to accept the design. Note that in the absence of a stated time scale for acceptance of design, the "period for reply" will apply. Many Project Managers are concerned that they have to "approve" the Contractor's design and therefore they are concerned as to whether they would be qualified, experienced and also insured to be able to do so.

Note the use of the word "acceptance" as distinct from "approval". Acceptance denotes compliance with the Scope or the applicable law, it does not denote that the design will work, that it will be approved by regulating authorities, or that it will fulfil all the obligations that the contract and the law impose; therefore performance requirements such as structural strength, insulation qualities, etc. do not need to be considered.

A reason for the Project Manager not accepting the Contractor's design is that it does not comply with the Scope or the applicable law. The Contractor cannot proceed with the relevant work until the Project Manager has accepted the design.

Note that under Clause 14.1, the Project Manager's acceptance does not change the Contractor's responsibility to Provide the Works or his liability for his design. This is an important aspect as, if the Project Manager accepts the Contractor's design but following acceptance there are problems with the proposed design, for example in meeting the appropriate legislation or the requirements of external regulatory bodies, this is the Contractor's liability.

Under Clause 27.1, the Contractor obtains approval of his design from Others where necessary. This will include planning authorities, and other third party regulatory bodies.

Note also, Clause 60.1(1), as above, second bullet point, which clearly refers to the fact that the Client's Scope prevails over the Contractor's Scope. This clause gives precedence to the Scope in Part 1 of the Contract Data over the Scope in Part 2 of the Contract Data. Thus the Contractor should ensure that the Scope he prepares and submits with his tender as Part 2 of the Contract Data, complies with the requirements of the Scope in Part 1 of the Contract Data.

Question 0.18 What does the term "Working Areas" mean in an NEC4 Engineering and Construction Contract? Presumably this is simply, the Site?

The term "Working Areas" is assumed by many NEC4 users to mean "the Site", but it is actually a far wider definition than "the Site" and needs a little explanation as confusion often arises.

The contract defines the Working Areas under Clause 11.2(20) as those parts of the working areas that are:

- necessary for Providing the Works and
- used only for work in this contract.

unless later changed in accordance with the contract.

The Working Areas are initially the Site, the boundaries of which are defined in Contract Data Part 1, but also any additional Working Areas may be identified by the Contractor in Contract Data Part 2 and submitted as part of his tender.

An example of an addition to the Working Areas could be where the footprint of the project to be built fills the Site, in which case the Contractor may name an additional area such as an adjacent field that he proposes to use for storage or to locate his Site compound.

In doing so, in order to qualify as an addition to the Working Areas, the Contractor should comply with Clause 11.2(20) in that this additional area is necessary for Providing the Works, and is used only for work in this contract.

Under Clause 16.3, the Contractor may also submit a proposal to the Project Manager for adding to the Working Areas during the carrying out of the works. The Project Manager may refuse acceptance because the Contractor has not complied with Clause 11.2(20).

The implications of adding to the Working Areas depend on the Main Option chosen.

Options A and B

The Short Schedule of Cost Components refers to the cost of resources used within the Working Areas, so these resources if also used within the extended Working Areas would be included as cost rather than within the Fee percentage when assessing compensation events.

Options C, D and E

The Schedule of Cost Components refers to the cost of resources used within the Working Areas, so these resources if also used within the extended working areas would be included as cost rather than within the fee percentage when assessing payments and compensation events.

Question 0.19 How do the NEC4 contracts deal with verbal communications? What do the words "in a form which can be read, copied and recorded" mean?

The NEC contracts have strict rules regarding communications under the contract, for example Clause 13.1 of the Engineering and Construction Contract states:

Each communication which the contract requires is communicated in a form which can be read, copied and recorded.

The terms "read, copied and recorded" can include communications sent by electronic means, for example email, or via a project intranet, so the issue and receipt are simultaneous.

Normally the Parties agree a protocol for project communications and NEC4 recognises this by referring in Clause 13.2 to a communication system as specified within the Scope.

This requirement for a communication to be in a form that can be read, copied and recorded is particularly important in respect of instructions, as the contract does not recognise verbal instructions; and whereas under other forms of contract a verbal instruction can be confirmed by the Contractor and, if not dissented from, normally within 7 or 14 days, become an instruction, the NEC4 contracts do not have that provision.

The instruction may be in the form of a letter, a pro forma instruction, or an email.

The Project Manager, Supervisor or Contractor replies within the period for reply stated in Contract Data Part 1, unless the contract states otherwise.

This period can be extended by mutual agreement between the communicating Parties.

If the Project Manager is required to accept or not accept, he must state his reasons for non-acceptance. Withholding acceptance for a reason stated in the contract is not a compensation event; this is particularly in respect of:

1   acceptance of Contractor's design;
2   acceptance of a proposed Subcontractor;
3   acceptance of a Contractor's programme.

The Project Manager issues his certificates (Payment, Completion and Termination Certificates) to the Client and to the Contractor, the Supervisor issues the Defects Certificate to the Project Manager, the Client and the Contractor.

Under Clause Notifications and certificates should be communicated separately from other communications; so for example, an early warning notice should not be included as part of a set of meeting minutes, or a letter that relates to other subjects.

Question 0.20   We wish to name Subcontractors that the Contractor is required to use in an NEC4 Engineering and Construction Contract. How can we do that?

When Clients are considering including and naming specialist Subcontractors in contracts they can normally do this by "nominating" them or by "naming" them.

Nominating Subcontractors

Although many other contracts provide for the inclusion of Prime Cost Sums within the tender documents in order that Nominated Subcontractors, chosen by the Client or his representatives, can be brought into the contract later by instructing the Contractor to use them, the NEC contracts have never done so, for a number of reasons:

1. The Contractor should be responsible for managing all that he has contracted to do. Nomination will often split responsibilities between the Contractor and the Client.
2. The use of Prime Cost Sums in Bills of Quantities means that the Contractor does not have to price that element of the work other than for profit and attendances; the more Prime Cost Sums, the less the tenderers have to price, and thus there is less pricing competition between the tenderers.
3. Contracts will often give the Contractor relief in the form of an extension of time in the event of a default by the Nominated Subcontractor, provided that the Contractor has done all that would be reasonable or practicable to manage the default.
4. Contracts will often give the Contractor relief in the form of loss and expense where it cannot be recovered from the Nominated Subcontractor, again provided the Contractor had done all that would be reasonable or practicable to manage the default. This would include the insolvency of a Nominated Subcontractor and a subsequent renomination.

Naming Subcontractors

In an NEC4 Engineering and Construction Contract, specialist Subcontractors could be named by the Client within the Scope, or information within the Scope may be limiting in that there is only one Subcontractor who could comply with that part of the Scope. It is important to recognise that if the Client names a specific Subcontractor, that Subcontractor will become solely the Contractor's responsibility and liability, as if the Contractor had chosen them himself.

Question 0.21 What happens under an NEC4 Engineering and Construction Contract Option B (priced contract with bill of quantities) if something is clearly indicated on the tender drawings, but it is missing from the Bill of Quantities?

It is a relatively simple question. Let us say you are tendering to build a new sports stadium under the NEC4 Engineering and Construction Contract Option B (priced contract with bill of quantities), and the drawings clearly show lighting to be installed under a roof. However, the Bill of Quantities make no reference to the lighting.

As a tendering Contractor, if you notice this omission from the Bill of Quantities at tender stage, would you, or should you, programme or price the roof? The initial answer of course is to raise a query during the tender period and get clarification – this at the very least will help your credibility and relationship with the potential Client.

There are a few contractual facts that we can deal with here:

- Information in the Bill of Quantities is not Scope (Clause 56.1) (or Site Information); it does not tell the Contractor what it has to do to price the works, it is simply a pricing document, which is also used for payment.

By the very nature of Option B, the risk in producing and verifying the Bill of Quantities lies with the Client (as opposed to Option A, where the risk in missing something from the activity schedule lies with the Contractor).

Clause 20.1 requires the Contractor to Provide the Works in accordance with the Scope.

Under Clause 17.1, if there is an ambiguity or inconsistency in or between contract documents, the Project Manager should give an instruction to resolve the matter.

- Clause 60.6 states that the Project Manager corrects mistakes in the Bill of Quantities which are due to ambiguities or inconsistencies – it also states that each such event is a compensation event.
- Clause 60.7 states that in assessing a compensation event that results from correction of an inconsistency between the Bill of Quantities and another contract document, the Contractor is assumed to have taken the Bill of Quantities as correct.

Taking the above points into account, the Contractor on an Option B contract would not have priced the lighting, as there is nothing to price within the Bills of Quantities.

It is probably debatable as to whether the Contractor would have programmed for the lighting as the programme would be based on the drawings and specifications (the Scope) rather than solely on the Bills of Quantities.

This element of the works (if required) will be assessed as a compensation event, and will include for the cost, and if necessary the time to include this element.

This leads to the following important conclusions.

As soon as tenderers notice something ambiguous or inconsistent, they should raise a tender enquiry. Tenderers may think they are giving up a potential commercial advantage, but it will give them real credibility during the tender and demonstrate their commitment to working in a "spirit of mutual trust and co-operation", as well as avoiding a contractual or political dispute on the project.

Notwithstanding the above, tenderers should price the Bill of Quantities, not the Scope.

Introduction to the NEC4 Contracts 59

Any omission from the Bill of Quantities will be assessed as a compensation event and Contractors are entitled to recover the financial effect of the change as well as the time effect, if planned Completion date is delayed.

Question 0.22 In an NEC4 Engineering and Construction Contract Option E (cost reimbursable contract), how are insurance premiums recovered by the Contractor?

Many Engineering and Construction Contract users believe that the cost of insurance premiums is included as Defined Cost within the Schedules of Cost Components, but this is not true.

The Schedule of Cost Components and the Short Schedule of Cost Components both state under Item 8.

The following are deducted from cost:

- the cost of events for which this contract requires the Contractor to insure and
- other costs paid to the Contractor by insurers.

Therefore:

- in the first bullet above, if a cost is incurred for which the Contractor is or was required to insure, then the Contractor cannot recover it as Defined Cost as it is deducted from Defined Cost.
- in the second bullet above, if the Contractor receives payments from the insurers, then these costs are deducted from Defined Cost.

So, how and where are the insurance premiums covered within the contract? . . . the answer is within the percentage for Fee.

This will also apply to insurance based costs such as the cost of ultimate holding company guarantees (Option X4), and performance bonds (Option X13).

Question 0.23 We wish to include for price fluctuations within an NEC4 Engineering and Construction Contract. How can we do that?

The inclusion of price fluctuations can be done through Secondary Option X1.

The Client should make the decision at the time of preparing the tender documents as to whether inflation for the duration of the contract is to be:

- The Contractor's risk – in which case it *should not* select Option X1
- The Client's risk – in which case, it *should* select Option X1.

The default within the ECC is that the contract is "fixed price" in terms of inflation, i.e. the Contractor has priced the work to include any inflation it may encounter during the period of the contract.

If Option X1 is chosen, the Prices are adjusted for inflation as the work progresses, by means of a formula.

Note that under Options C and D, whether Option X1 is selected or not, the Price for Work Done to Date is the current cost at the time it is incurred. Option X1 is then applied to the Total of the Prices (the target).

Option X1 is not applicable to Options E and F as the Client again pays Defined Cost at the time that it is incurred.

The key components of the formula are:

- The "Base Date Index" (B) is the latest before the Base Date.
- The "Latest Index" (L) is the latest available index before the assessment date of an amount due.
- The "Price Adjustment Factor" is the total of the products of each of the proportions stated in the Contract Data multiplied by (L – B)/B for the index linked to it.

Options A and B

Under Options A and B, the amount due includes an amount for price adjustment which is the sum of:

- the change in the Price for Work Done to Date since the last assessment of the amount due multiplied by the PAF and
- the amount for price adjustment included in the previous amount due.

The change in the Price for Work Done to Date = £50,000

The Base Date Index (B) = 280.0

The Latest Index (L) = 295.5

The Price Adjustment Factor is therefore (L – B)/B

= (295.5 – 280.0)/280.0

= 0.055

Inflation since the base date is therefore 5.5%

The amount due in this assessment is therefore

£50,000.00 × 0.055 = **£2,750.00**

Under Options C and D, the amount due includes an amount for price adjustment which is the sum of:

- the change in the Price for Work Done to Date since the last assessment of the amount due multiplied by (PAF/(1+PAF)) where PAF is the Price Adjustment Factor for the date of the current assessment and
- correcting amounts, not included elsewhere, which arise from changes to indices used for assessing previous amounts for price adjustment.

This amount is then added to the Total of the Prices, i.e. the target.

Note that if Option X1 is chosen, then Defined Cost for compensation events is assessed by adjusting current Defined Cost back to the base date.

Note that for compensation events the Defined Cost is assessed using
- the Defined Cost at base date levels for amounts in the Contract Data for people and Equipment and
- the Defined Cost current at the dividing date used in assessing the compensation events adjusted to the base date by dividing by "one plus the PAF" for the last assessment due before that dividing date for other amounts.

Aside from Clients favouring fixed price contracts in terms of contractual risk, the quite complex rules of Option X1 are probably one of the reasons why the Option is rarely chosen!

Question 0.24 A new legal requirement has just come into force. The Contractor on an NEC4 Engineering and Construction Contract Option C (target contract with activity schedule) has stated that he is entitled to additional payment to recover this. Is he correct?

As with Option X1 above, the default is that the contract is "fixed price" in terms of changes in the law, i.e. the Contractor has priced the work to include any changes in the law it may encounter during the period of the contract.

If Option X2 is chosen, and a change in the law occurs after the Contract Date it is a compensation event. The Prices may be increased or reduced in addition to providing for any delay to Completion.

Note that Option X2 refers to a change in the law of the country in which the Site is located, so, for example, a change in the law in another country where goods are being fabricated for delivery to the Site will not be a compensation event.

Question 0.25 We wish to enter into a partnering arrangement with the Contractor within an NEC4 Engineering and Construction Contract. What can we do in order to achieve this?

The answer to the question really depends on what you mean by "partnering arrangement".

The past 25 years, particularly since the publication of the Latham and Egan reports, have seen the growth of partnering and framework agreements.

The US Construction Industry Institute has defined partnering as:

> *A long term commitment between two or more organizations for the purpose of achieving specific business objectives by maximizing the effectiveness of each participant's resources . . . the relationship is based upon trust, dedication to common goals and an understanding of each other's individual expectations and values.*

Partnering is a medium to long-term relationship between contracting Parties, whereby the Contractor, and in turn Consultants and various other Parties, are not required to tender competitively on price for each project, but are awarded the work by the Client on the basis of their ability to deliver, and normally the price is fixed by negotiation.

The construction industry is known to be a high risk business, and many projects can suffer unexpected cost and time overruns frequently resulting in disputes between the Parties. The risks within a project are initially owned by the Client, who may choose to adopt a "risk transfer" approach where the risks are assigned through the contract to the Contractor who has the opportunity to price and programme for them, or a "risk embrace" approach where the Client retains the risks. In reality, most contracts are a combination of the two.

The traditional approach to risk management is that of risk transfer, which is fine if the scope of work is clear and well defined; however, in recent years Clients have become more aware that they can achieve their objectives better by adopting a more "old-fashioned" risk embrace culture.

Advantages of partnering agreements are:

- The procurement process from conception to commencement on Site, and subsequent completion, can be significantly reduced in terms of time and resources.
- No re-tendering time and cost for future projects.
- Relationships can be developed based on trust.
- Contractors are appointed earlier and can contribute to the design and procurement process through ECI (Early Contractor Involvement).
- There tends to be greater cost certainty.
- Continuous improvement can be achieved by transferring learning from one project to the next.
- Better working relationships can be developed as the Parties know each other.
- Prices can be more competitive and resources used more efficiently by continuous flows of work.

Disadvantages of partnering agreements are:

- Obligations and liabilities can become less clear in time as the Parties can lapse into informal working patterns.
- Complacency can set in after some time as the Contractor has an assured flow of work.
- Clients are often unsatisfied that they are getting value for money in their projects.
- Continuing to award the work to a small number, or even one Contractor, prevents other equally as good, or better Contractors having the opportunity to carry out work.

All of these disadvantages can be overcome by maintaining a disciplined approach to communications between the Parties and their rights and obligations, and also by continuous measurement of performance and deliverables.

There are two methods of entering into a partnering arrangement between the Client and the Contractor and with other members of the supply chain:

1   Secondary Option X12 (multi-party collaboration)

Secondary Option X12 was formerly in NEC3 called "*Partnering*", enabling a multi-party partnering agreement to be implemented between various Parties.

In this case Option X12 is used as a Secondary Option common to the contract which each party has with the body that is paying for the work.

The content of Option X12 is derived from the "Guide to Project Team Partnering" published by the Construction Industry Council (CIC). It is estimated that Option X12 is used on less than 10% of NEC3 contracts.

It must be stressed that no legal entity is created between the Partners, so it is not a partnership as such.

Some definitions need to be explained:

(i)   The Partners are those named in the Schedule of Partners.
(i)   An Own Contract is a contract between two Partners.
(ii) The Core Group comprises the Partners selected to take decisions on behalf of the Partners.
(iii) The Schedule of Core Group Members is a list of Partners forming the Core Group.
(iv) Partnering Information is information which specifies how the Partners work together.
(v)   A Key Performance Indicator is an aspect of performance for which a target is attached in the Schedule of Partners.

Each Partner, represented by a single representative, is required to work with the other Partners in accordance with the Partnering Information to achieve the

Promoter's (normally their joint Client) objective stated in the Contract Data and the objectives of every other Partner.

The Core Group acts and takes decisions on behalf of the Partners. The Core Group also keeps up to date a Schedule of Core Group Members and a Schedule of Partners.

The Partners are required to work together, using common information systems, and a Partner may ask another Partner to provide information that it needs to carry out the work in its Own Contract.

The Core Group may give an instruction to the Partners to change the Partnering Information, which is a compensation event.

The Core Group also maintain a timetable showing the Partners' contributions. If the Contractor needs to change its programme to comply with the timetable then it is a compensation event.

Each Partner also gives advice, information and opinion to the Core Group where required.

Each Partner must also notify the Core Group before subcontracting any work, though it does not say that the Core Group is required to respond to the notification.

Finally Option X12 provides for Key Performance Indicators (KPIs) with amounts paid as stated in the Schedule of Partners. The Promoter may add a KPI to the Schedule of Partners but cannot delete or reduce a payment.

2   NEC4 Alliance Contract

The NEC4 Alliance Contract is different from the others within the NEC4 family of contracts as it is a multi-party contract with an integrated risk and reward model.

The Alliance Contract creates a "true" alliance arrangement in which the Client and all key members of the supply chain, called "Partners", are engaged under a single contract.

The basis of the Alliance Contract will be that all Parties will work together in achieving the Client objectives, and share in the risks and benefits of doing so.

It is a multi-party contract, under which the Client and its main delivery partners, which potentially consists of Works Contractors, Consultants and Equipment Suppliers, all sign up to a single set of terms.

The members of the alliance are referred to within the contract as the "Alliance Delivery Team", a single integrated delivery structure.

The form is effectively a reimbursable contract, equivalent to NEC Option E; overlaid with a performance regime that consists of targets for meeting alliance objectives set out in a performance table.

A project budget is measured against the total alliance cost (similar in nature to the pain–gain approach of NEC option C) to determine the "pain–gain" share. The critical difference is that the calculation of the budget and total alliance cost includes the client's costs, as the client is an integral part of the alliance. Similar to NEC option E, the alliance member's costs consist of the alliance member's defined cost plus fee.

The contract is designed for use on major projects or programmes of work, where longer term collaborative ways of working are to be created. It can also be used to deliver a programme where a number of lower-value projects can be combined to create a major programme of work.

Further collaboration can be achieved through use of Secondary Option X12, "multiparty collaboration", which incentivises multiple suppliers to collaborate to achieve a common set of objectives set by the promotor (commonly the Client).

The Alliance Contract takes Option X12 and builds this into the core of the contract, creating the requirement for members of the alliance to collaborate with each other to achieve alliance objectives and partner objectives. To achieve this, alliance members work collectively to support delivery of the contract and establish an integrated alliance delivery team on a best-for-project basis.

The Client has a dual role in that it has certain retained powers and functions that it performs outside of the alliance, as well as the power and functions of an alliance member.

The alliance board has overall responsibility for the alliance and sets strategy, appoints the alliance manager, makes decisions and resolves disputes. Each alliance member has an alliance board representative, including the client.

The alliance manager manages the contract on behalf of the alliance and undertakes many of the functions exercised by the Project Manager or service manager under other NEC contracts, as well as some aspects of the Contractor's role.

Reflecting the collaborative nature of the contract, most alliance decisions have to be made unanimously by the alliance board. Alliance members share the majority of risk under the contract and agree that there can be no claims made against other members of the alliance except for very limited exceptions, principally due to a deliberate breach of contract.

Payment by the Client to the Partners is on the basis of Defined Cost and all Partners are incentivised to achieve alliance objectives through a performance table.

The performance table sets out what performance is required and the reward or deduction regimes that apply if the performance targets are over or under achieved. It also includes an assessment of pain or gain share if the alliance's costs, which includes the client's costs as well as the defined cost plus fee of the partners, is above or below the budget.

The structure of the NEC4 Alliance Contract

This contract is designed for use on major projects or programmes of work where longer term collaborative ways of working are to be created.

Core clauses

1 General
2 The Alliance's main responsibilities

3   Time
4   Quality management
5   Payment
6   Compensation events
7   Title
8   Liabilities and insurance
9   Termination, resolving and avoiding disputes.

Secondary Option clauses

Option X2	Changes in the law
Option X4	Performance guarantee
Option X9	Transfer of rights
Option X10	Information modelling
Option X18	Limitation of liability
Option X22	Early Alliance involvement
Option X26	Programme of work
Option Y(UK)1	Project Bank Account
Option Y(UK)2	The Housing Grants, Construction and Regeneration Act 1996
Option Y(UK)3	The Contracts (Rights of Third Parties) Act 1999
Option Z	Additional conditions of contract.

Schedule of Cost Components

Contract data

Part 1 – Client and Alliance Data

Part 2 – Data provided by each partner

Although the NEC suite discourages the use of "Z clauses", it is likely that additional client-led provisions may be necessary dependent on what is to be provided.

As the equivalent of a design and build contract, the intention is that the Alliance Contract will be entered into at the same stage of project maturity as the NEC Engineering and Construction Contract. There is, however, an Early Alliance Involvement option X22 allowing for engagement between the Parties at a much earlier stage – for example, allowing early design studies to be undertaken and the project budget to be developed. The client can add new schemes to the alliance using Option X26 (Programme of work).

In keeping with the ethos of shared risk, the contract includes a provision that encourages Parties to avoid adopting adversarial positions. The Alliance Board is intended to play a key role in resolving disputes. An independent expert can be called on to give a non-binding opinion and senior representatives must play an active role, supported by mediation if agreed.

Clearly, the statutory right to adjudicate cannot be excluded but, unlike in other NEC forms, does not facilitate adjudication by way of any extended contractual provisions. The expectation is that the Scheme for Construction Contracts would apply.

Question 0.26 Within the NEC4 Engineering and Construction Contract, when would we use Option X15 (The Contractor's Design)?

This is an Option, formerly in NEC3 called *"Limitation of the Contractor's liability for his design to reasonable skill and care"*.

There are several items covered by this Option.

First, whilst the Scope within the NEC4 Engineering and Construction Contract defines what, if any, design is to be carried out by the Contractor, the contract is silent on the standard of care to be exercised by the Contractor when carrying out any design.

Two terms that relate to design liability are "fitness for purpose" and reasonable skill and care".

Whilst this book is intended for international use, in defining the term "fitness for purpose" one must look in English law primarily to the Sale of Goods Act 1979, which refers to the quality of goods supplied including their state and condition complying in terms of "fitness for all the purposes for which goods of the kind in question are commonly supplied".

In construction, fitness for purpose means producing a finished project fit in all respects for its intended purpose. This is an absolute duty independent of negligence, and in the absence of any express terms within the contract to the contrary, a Contractor who has a design responsibility will be required to design and build the project "fit for purpose".

Some contracts will limit the Contractor's liability to that of a consultant, i.e. reasonable skill and care.

The Engineering and Construction Contract does this through Option X15. If this Secondary Option is not chosen, the Contractor's liability for design is "fitness for purpose". If the Contractor corrects a Defect for which it is not liable, it is a compensation event.

Contrast this with the level of liability of a consultant (whether acting for a Client or a Contractor) in providing a design service, which is defined, again in English law by the Supply of Goods and Services Act 1982, where there is an implied term that the consultants will carry out the service, in this case design, with reasonable care and skill, which means designing to the level of an ordinary, but competent person exercising a particular skill.

So, in the absence of any express terms to the contrary within the contract, a designer will normally be required to design using "reasonable skill and care". This is normally achieved by the designer following accepted practice and complying with national standards, codes of practice, etc.

Clearly, despite the Contractor believing that it has offset its design obligations and liability to its designing consultant, it has to be aware that it, and its consultant have differing levels of care and liability.

The remaining items covered by this Option are:

- The Contractor may use material provided to it under the contract, unless the ownership of the material has been given to the Client.
- The Contractor retains copies of drawings, specifications, reports, etc. in the form stated in the Scope for the period of retention, normally six or twelve years after Completion of the works.
- The Contractor may be required to provide Professional Indemnity (PI) Insurance if required, and in the amount in the Contract Data.

Question 0.27 **We note that the NEC4 Engineering and Construction Contract includes Option X17 (low performance damages). What is this, and how is it calculated and paid to the Client?**

In the event that the Contractor produces defective work, the Client has three options:

(i) The Contractor corrects the Defect (Clause 44.1).
(ii) If the Contractor does not correct the Defect, the Project Manager assesses the cost to the Client of having the Defect corrected by other people and the Contractor pays this amount (Clause 46.1).
(iii) The Client can accept the Defect and a quotation from the Contractor for reduced Prices and/or an earlier Completion Date (Clause 45).

Where the performance in use fails to reach the specified level within the contract, and the Contractor cannot or will not correct the Defect, the Client can take action against the Contractor to recover any damages suffered as a result of the breach, but as an alternative can recover low performance damages under Option X17 if it has been selected.

Example

The Scope requires the Contractor to design and install a HVAC system to a major retail development.

There are specific and measurable performance criteria for the system including temperature variations and energy efficiency. The system is required to have a design life of 30 years (360 months).

> The Scope states that the system will be tested at Completion and includes a table showing how the performance of the system will be measured and acceptable levels of achievement.
>
> The system is expected to perform to 98–100% of performance criteria. If it falls within 90–98% the system will be accepted, but low performance damages will be payable by the Contractor to the Client. If it falls below 90% it will not be accepted.
>
> Contract Data Part 1 contains the following entries:
>
> | amount per month | performance level |
> | --- | --- |
> | £50.00 | 96% – 98% |
> | £150.00 | 94% – 96% |
> | £200.00 | 92% – 94% |
> | £250.00 | 90% – 92% |
>
> When tested, the system achieves 95% of the performance criteria.
>
> Therefore, 360 months × £150.00 = £54,000 is payable by the Contractor at the Defects Date.

Question 0.28 We do not understand NEC4 Engineering and Construction Option X18 (limitation of liability). When would this be applied?

If the Contractor causes any loss of or damage to the Client or his property, it would normally be liable to the Client for the full cost of remedial works; however, Clause X18 provides for this liability to be limited to amounts stated in the Contract Data.

The Contractor's liability to the Client for the Client's

- indirect or consequential loss
- loss or damage to its property

may be limited to amounts stated in the Contract Data.

In addition, the Contractor's liability to the Client for latent defects due to its design may again be limited to amounts stated in the Contract Data.

Clause X18.5 can be used to place limits on the total liability the Contractor has to the Client for all matters under the contract other than excluded matters in contract, tort or delict.

Excluded matters are:

- loss of or damage to the Client's property
- delay damages if Option X7 applies
- low performance damages if Option X17 applies
- Contractor's share if Option C or D applies.

The Contractor is not liable for any matter unless it has been notified to the Contractor before the end of the liability date which is stated in the Contract Data in terms of years after the Completion of the whole of the works.

In the UK, this may be six or twelve years dependent on the type of contract; other legislations set this at ten years.

Question 0.29 We wish to include for Key Performance Indicators (KPIs) with an NEC4 Engineering and Construction Contract? How can we do this?

Performance of the Contractor can be monitored and measured against Key Performance Indicators (KPIs) using Option X20.

Targets may be stated for Key Performance Indicators in the Incentive Schedule.

From the starting date until the Defects Certificate is issued, the Contractor is required to report his performance against KPIs to the Project Manager at intervals stated in the Contract Data including the forecast final measurement. If the forecast final measurement will not achieve the target stated in the Incentive Schedule the Contractor is required to submit his proposals to the Project Manager for improving performance.

The Contractor is paid the amount stated in the Incentive Schedule if the target for a KPI is improved upon or achieved. Note that there is no payment due from the Contractor if he fails to achieve a stated target. The Client may add a new KPI and associated payment to the Incentive Schedule but may not delete or reduce a payment.

Question 0.30 How should Z clauses be incorporated into an NEC4 contract? Are there any recommended Z clauses?

Probably the best answer to this question is that Z clauses should be used "as sparingly as possible"!

Option Z allows conditions to be added to, or omitted from, the core clauses.

All changes to the core clauses should be included as Z clauses rather than amending the core clauses themselves, so in effect the clause remains in the contract, but is amended within the Z clause. It is also critical that, when drafting a

Z clause, it must be clearly stated what happens to the original core clause – for example, it is deleted. If the original core clause is not deleted, then it is likely that an inconsistency will arise which will be interpreted against the party who wrote or amended the clause, i.e. the Client.

These conditions may modify or add to the core clause, to suit any risk allocation or other special requirements of the particular contract. However, changes should be kept to a minimum, consistent with the objective of using industry standard, impartially written contracts.

Some NEC practitioners have stated that the use (or overuse) of Z clauses is damaging to the NEC contracts.

The author believes that it is not the use (or overuse) of Z clauses that is the primary problem, it is the incorrect drafting of Z clauses, or using a Z clause when the contract or a provision already within the contract is not understood and ambiguity arises, that is the problem.

It must be remembered that if a Client amends a contract to allocate a risk to the Contractor which may have been intended to be held by the Client, the Contractor has a right to price it in terms of time and money. Therefore, the practice of Clients amending contracts to pass risk to Contractors without considering who is best able to price, control and manage those risks can in many cases prove to be unwise and uneconomical.

It is important to spend time considering whether a Z clause is appropriate in each case, then when that decision has been made, that the clause is drafted correctly and aligned to the drafting principles of the original contract, in the case of the Engineering and Construction Contract using ordinary language, present tense, short sentences, bullet pointing and italicising of terms identified in the Contract Data. It is not unusual to see Z clauses in an Engineering and Construction Contract written in a legalistic language, in the future tense, with no italics, and without punctuation apart from full stops!

There are no recommended Z clauses, but some that tend to be used fairly commonly within the NEC4 Engineering and Construction Contract are:

(i) The Contractor's share
   Delete:
   Clause 54.3
   The *Project Manager* makes a preliminary assessment of the *Contractor's* share at Completion of the whole of the *works* using his forecasts of the final Price for Work Done to Date and the final total of the Prices. This share is included in the amount due following Completion of the whole of the *works*.
   Add:
   New Clause in lieu
   The *Project Manager* makes a preliminary assessment of the *Contractor's* share at any assessment date.

- If the forecast Price for Work Done to Date is less than the final total of the Prices, this share is included in the amount due following Completion of the whole of the *works*.
- If the forecast Price for Work Done to Date is more than the forecast total of the Prices, this share is retained from the amount due.

Reason for Z clause:

Many Clients believe that the Contractor's share should be calculated when the Price for Work Done to Date reaches the current Total of the Prices (the target) rather than on Completion, so that the Contractor is not paid monies that it will have to pay back later.

(ii) For Options C, D, E & F

Delete:

Clause 11.2(31)

The Price for Work Done to Date is the total Defined Cost which the *Project Manager* forecasts will have been paid by the *Contractor* before the next assessment date plus the Fee.

Add:

New Clause in lieu:

The Price for Work Done to Date is the total Defined Cost which the Contractor has paid plus the Fee.

Reason for Z clause:

Many Clients do not accept Clause 11.2(31) in that a forecast has to be made of the Defined Cost up to the *next* assessment date and that forecast included within the payment. The amended clause aligns with pre NEC3 contracts.

(iii) Testing and Defects

Searching for and notifying Defects

Delete:

Clause 43.2

Until the *defects date*, the *Supervisor* and the *Contractor* notify the other as soon as they become aware of a Defect.

Add:

Until the *defects date*, the *Supervisor* notifies the *Contractor* of each Defect as soon as he finds it.

Reason for Z clause:

Why does the Supervisor need to know the Contractor has found a Defect? Surely, if he needs to know about a Defect, he only needs to know that the Contractor has corrected that Defect?

Question 0.31 We are confused by the fact that within the NEC4 Engineering and Construction Contract, there is the Schedule of Cost Components and the Short Schedule of Cost Components? What are these for and what is the difference between the two?

There are two Schedules of Cost Components in the contract.

This is because there are differing uses for Defined Cost, dependent on which main Option is chosen as follows:

> Option A Defined Cost is only used for assessing compensation events
>
> Option B

therefore the **Short Schedule of Cost Components** is used

> Option C Defined Cost is used for assessing compensation events, and for assessing Price for Work Done to Date.
>
> Option D
>
> Option E

therefore the **Schedule of Cost Components** is used, though the Short Schedule of Cost Components can be used by agreement between the Project Manager and the Contractor for assessing compensation events.

> Option F Defined Cost is the amount of payment due to Subcontractors and the prices for work done by the Contractor itself.

Therefore, neither the Schedule of Cost Components nor Short Schedule of Cost Components is used.

The Schedule of Cost Components

The Schedule of Cost Components only applies when Option C, D or E is used.

Clause 1 – People

This relates to the cost of people who are directly employed by the Contractor and whose normal place of working is within the Working Areas, i.e. the Site and any another area named by the Contractor in Contract Data Part 2, and any further additions subsequently agreed under Clause 16.3. Also people whose normal place of working is not within the Working Areas but who are working in the Working Areas proportionate to the time they spend working within the Working Areas, i.e. people who are based off site, but who are on site temporarily.

The cost component covers the full cost of employing the people including wages and salaries paid whilst they are in the Working Areas, payments made to the people for bonuses, overtime, sickness and holiday pay, special allowances, and also payments made in relation to people for travel, subsistence, relocation, medical costs, protective clothing, meeting the requirements of the law, a vehicle and safety training.

The Schedule also relates to the cost of people who are not directly employed by the Contractor but are according to the time worked while they are in the Working Areas.

Examples would be agency labour, cleaners or security guards, in all cases paid by the hour rather than on a lump sum price basis.

The Schedule of Cost Components does not provide for people rates to be priced as part of the Contractor's tender, the principle being that the People component is real cost rather than a rate that has been forecast at tender stage months or even years before it is required to be used. This is a fairer method of establishing cost in that the risk is not the Contractor's, but in practice requires the Contractor if necessary to prove the cost of each operative, tradesman and member of staff, which can at times be laborious.

For this reason, many clients adopt a more "traditional policy" of including a schedule of rates to be priced by the Contractor at tender stage, these rates then being used for payments and compensation events where appropriate. These rates are also referred to during the tender assessment process.

Clause 2 – Equipment

This relates to the cost of Equipment (referred to as "Plant" in other contracts) used within the Working Areas. If the Equipment is used outside the Working Areas, then it is deemed to be included in the fee percentage.

If the Equipment is hired externally by the Contractor, the hire rate is multiplied by the time the Equipment is required. The cost of transport of Equipment to and from the Working Areas and any erection and dismantling costs are also separately costed.

If the Equipment is owned by the Contractor, or hired by the Contractor from a company within the parent company such as an internal "plant hire" company, the cost is assessed at open market rates (not at the rate charged by the hirer) multiplied by the time for which the Equipment is required.

Difficulty has often arisen with past editions (pre NEC3) of the contract where a piece of Equipment is owned by the Contractor, or a company within the group, in which case there would be no invoice as such to prove the cost – in many cases it was just an internal charge or cost transfer.

In those previous editions, the drafters sought to deal with the problem by establishing the weekly cost by the use of a formula:

$$\frac{\text{Purchase price of the Equipment}}{\text{Working life remaining at purchase}} \times \text{Depreciation and maintenance \%}$$

Example

Purchase Price = £30,000

Working life remaining = 250 weeks

Depreciation and maintenance percentage = 20%

The weekly cost is:

$$\frac{£30,000}{250} \times 1.20 = £144.00 \text{ per week}$$

In theory, this should provide an equitable solution, but in practice it has provided some inequitable answers, and also it has to be applied to EVERY piece of Equipment owned by the Contractor! One hoped the Contractor did not have too many of its own wheelbarrows on the site!

In NEC3 the drafters then changed to "open market rates multiplied by the time for which the Equipment is required" and this has followed through into NEC4.

In truth major contractors who have a large "plant" company will benefit in this respect from advantageous discount agreements with their suppliers that are well below "open market rates".

Equipment purchased for use in the contract is paid on the basis of its change in value (the difference between its purchase price and its sale price at the end of the period for which it is used) and the time related on cost charge stated in Contract Data Part 2 for the period the Equipment is required.

During the course of the contract, the Contractor is paid the time related charge per time period (per week/per month) and when the change in value is determined, a final payment is made in the next assessment.

Example

The Contractor has purchased 6 Portakabins at £6,250.00 each for use on a project of 4 years duration. This cost includes supply and delivery of the Portakabins.

- Total purchase price = 6 × £6,250.00 = £37,500.00
- On completion, the Portakabins are sold for £2,750.00 each, therefore the total sale price is £16,500.00
- The change in value is therefore £37,500.00 – £16,500.00 = £21,000.00.

Any special Equipment is paid on the basis of its entry in Contract Data Part 2.

Consumables such as fuel are also separately costed including any materials used to construct or fabricate Equipment.

The cost of transporting Equipment to and from the Working Areas and the erection, dismantling and any modifying of the Equipment is costed separately.

Note: Any People cost such as Equipment drivers and operatives involved with erection and dismantling, or use of Equipment should be included in the cost of People, *not* the Equipment they work on.

Clause 3 – Plant and Materials

This deals with purchasing Plant and Materials, including delivery, providing and removing packaging and any necessary samples and tests.

The cost of disposal of Plant and Materials is credited.

Clause 4 – Subcontractors

This covers payments to Subcontractors for work that is subcontracted, without taking into account amounts paid to or retained from the Subcontractor which would result in the Client paying or retaining the same amount twice; for example if the Contractor deducts retention from a Subcontractor's payment, the amount before retention is paid, then the Client may separately deduct retention from the Contractor.

Clause 5 – Charges

This covers various miscellaneous costs incurred by the Contractor such as temporary water, gas and electricity, payments to public authorities, and also payments for various other charges such as cancellation charges, buying or leasing of land, inspection certificates and facilities for visits to the Working Areas.

The cost of any consumables and equipment provided by the Contractor for the Project Manager's or Supervisor's offices is also included as direct cost.

Note that under a new provision within the NEC4 Engineering and Construction Contract, the cost of the Contractor's own consumables (referred to by the author in his numerous NEC training courses as "tea bags and toilet rolls"!) are not included within this cost component, and therefore unless they are in another cost component, and it is difficult to see where the costs have been relocated, they are deemed to be included within the fee percentage.

This is a change from previous editions of the Engineering and Construction Contract, where the cost is calculated by multiplying the Working Areas overheads percentage inserted by the Contractor in Contract Data Part 2 by the People cost items.

Clause 6 – Manufacture and fabrication

This relates to the components of cost of manufacture or fabrication of Plant and Materials outside the Working Areas. Hourly rates are stated in Contract Data Part 2 for the categories of employees listed.

Clause 7 – Design

This deals with the cost of design outside the Working Areas. Again, hourly rates are stated in Contract Data Part 2 for the categories of employees listed.

Clause 8 – Insurance

The cost of events for which the Contractor is required to insure and other costs to be paid to the Contractor by insurers are deducted from cost.

The Short Schedule of Cost Components

The Short Schedule of Cost Components is restricted to the assessment of compensation events under Option A and B.

Clause 1 – People

This relates to the cost of people who are directly employed by the Contractor and whose normal place of working is within the Working Areas, i.e. the Site and any another area named by the Contractor in Contract Data Part 2, and any further additions subsequently agreed under Clause 16.3. Also people whose normal place of working is not within the Working Areas but who are working in the Working Areas proportionate to the time they spend working within the Working Areas, i.e. people who are based off site, but who are on site temporarily.

The amounts for People costs are calculated by multiplying the People Rates by the total time appropriate to that rate spent within the Working Areas. This is a new provision within the NEC4 Engineering and Construction Contract.

If there is no People Rate for a specific category of person, then the Project Manager and the Contractor can agree a new rate.

Clause 2 – Equipment

This relates to the cost of Equipment used within the Working Areas.

The cost of Equipment is calculated by reference to a published list, for example BCIS (Building Cost Information Service) Schedule of Basic Plant Charges or the CECA (Civil Engineering Contractors Association) Dayworks Schedule. In Contract Data Part 2, the Contractor names the published list to be used and also inserts a percentage for adjustment against items of Equipment in the published list.

The Contractor also inserts rates into Contract Data Part 2 for Equipment not included within the published list. Any Equipment required that is not in the published list or priced within Contract Data Part 2 is then priced at competitively tendered market rates.

The time the Equipment is used is as referred to in the published list, which may have hourly, weekly or monthly rates.

By referring to the published list, whether the Equipment is then owned or hired by the Contractor is irrelevant.

The cost of transporting Equipment to and from the Working Areas and the erection and dismantling of the Equipment is costed separately if not included within the published list. The cost of Equipment operators is included within the People costs.

Any Equipment not included within the published lists is priced at competitively tendered or open market rates.

Clause 3 - Plant and Materials

This is the same as for the Schedule of Cost Components and also deals with purchasing, delivery, providing and removing packaging and any necessary samples and tests. The cost of disposal of Plant and Materials is credited.

Clause 4 - Subcontractors

This covers payments to Subcontractors for work that is subcontracted.

Clause 5 - Charges

This covers various miscellaneous costs incurred by the Contractor such as temporary water, gas and electricity payments to public authorities, and also payments for various other charges, which may or may not be relevant dependent on the project.

Clause 6 - Manufacture and fabrication

This relates to the components of cost of manufacture or fabrication of Plant and Materials outside the Working Areas. The calculation is based on amounts paid by the Contractor.

Clause 7 - Design

This is exactly the same as for the Schedule of Cost Components and deals with the cost of design outside the Working Areas. Again, hourly rates are stated in Contract Data Part 2 for the categories of employees listed.

Clause 8 – Insurance

The cost of events for which the Contractor is required to insure and other costs to be paid to the Contractor by insurers are deducted from cost (se Table 0.1).

Table 0.1 Differences between the Schedule of Cost Components and the Short Schedule of Cost Components

Schedule of Cost Components	Short Schedule of Cost Components
People 1	
Detailed list of People costs classified as Defined Cost	No detailed list, People Rates multiplied by the total time spent within the Working Areas
Equipment 2	
Detailed list of Equipment costs classified as Defined Cost	Reference to the published list identified within Contract Data Part 2.
Plant and Materials 3	
No difference Payments for purchasing Plant and Materials, including delivery, samples and tests.	
Subcontractors 4	
Payments to Subcontractors without taking into account amounts paid to or retained which would result in the Client paying or retaining the amount twice.	Payments to Subcontractors
Charges 5	
No difference Detailed list of Charges classified as Defined Cost	
Manufacture and fabrication 6	
No difference Reference to hours worked multiplied by rates in Contract Data Part 2.	
Design 7	
No difference Reference to hours worked multiplied by rates in Contract Data Part 2.	
Insurance 8	
No difference Amounts deducted from cost where relevant.	

Question 0.32 How are Subcontractors dealt with within the NEC4 Engineering and Construction Contract Schedules of Cost Components?

The NEC3 Engineering and Construction Contract

Within the NEC3 Engineering and Construction Contract there were two Schedules of Cost Components, the Schedule of Cost Components and the Shorter Schedule of Cost Components.

- For Options A and B, the Shorter Schedule of Cost Components was a complete statement of the cost components under the definition of Defined Cost whether work was subcontracted or not, so it included Subcontractor costs.
- For Options C, D and E, the Schedule of Cost Components meant the Contractor and not his Subcontractors, the cost of Subcontractors having to be added to the cost covered by the Schedule of Cost Components to calculate the total Defined Cost.

So with both Schedules of Cost Components, there was no cost components specifically covering Subcontractor costs.

Note: The Schedules of Cost Components do not apply to Option F (Management Contract) where Defined Cost is the amount of payments due to Subcontractors for work that is subcontracted, less Disallowed Cost.

The NEC4 Engineering and Construction Contract

Within the NEC4 Engineering and Construction Contract there are again two Schedules of Cost Components, the Schedule of Cost Components and the Short Schedule of Cost Components.

In both cases the cost of Subcontractors is specifically identified:

- For Options A and B, the Short Schedule of Cost Components includes cost component 4 which refers to *"payments to Subcontractors for work which is subcontracted"*.
- For Options C, D and E the Schedule of Cost Components includes cost component 4 which refers to *"payments to Subcontractors for work which is subcontracted without taking into account any amounts paid to or retained from the Subcontractor by the Contractor, which would result in the Client paying or retaining the amount twice"*.

Note: Again, the Schedules of Cost Components do not apply to Option F (Management Contract) where Defined Cost is the amount of payments due to Subcontractors for work that is subcontracted, less Disallowed Cost.

Introduction to the NEC4 Contracts   81

Question 0.33   Some of the NEC4 Contracts include a Price List. What is this and how do we use it?

The larger NEC4 contracts include Main and Secondary Options allowing the user to select a procurement strategy and payment mechanism most appropriate to the project and the various risks involved; essentially, the Main Options differ in the way the Contractor is paid.

For example, the NEC4 Engineering and Construction Contract provides the following Main Options:

> Option A   Priced contract with activity schedule
> Option B   Priced contract with bill of quantities
> Option C   Target contract with activity schedule
> Option D   Target contract with bill of quantities
> Option E   Cost reimbursable contract
> Option F   Management contract.

- Options A and B are priced contracts in which the risks of being able to carry out the work at the agreed prices are largely borne by the Contractor.
- Options C and D are target contracts in which the Client and Contractor share the financial risks in an agreed proportion.
- Options E and F are two types of cost reimbursable contract in which the financial risks of being able to carry out the work are largely borne by the Client.

Note that the lettering of the main options is common for all the NEC4 contracts, so Option A is always a Priced Contract, Option C is always a Target Contract, Option E is always Cost Reimbursable, etc.

Some of the NEC4 Contracts, for example, the NEC4 Term Service Contract have provision for a Price List:

> Option A   Priced contract with price list
> Option C   Target contract with price list
> Option E   Cost reimbursable contract.

In addition, the NEC4 Short Contracts also provide for Price Lists, this time instead of Main Options.

For example, the Engineering and Construction Short Contract is used for "*contracts which do not require sophisticated management techniques, comprise straightforward work and impose only low risks on both the Client and the Contractor*", in which case, there is no need for a wide range of procurement strategies as provided in the NEC4 Engineering and Construction Contract and other NEC4 contracts, so the NEC4 Engineering and Construction Short Contract includes within the Contract Data a single page Price List.

The Price List

The Price List can be amended to suit particular Client and project/service requirements, but the templates within the contract are set out as follows.

The columns are typically headed:

1. Item Number
2. Description
3. Unit
4. Quantity
5. Rate
6. Price.

As the guidance note above the Price List states:

- Entries in the first four columns are made either by the Client or the tenderer (Contractor/Consultant), entries in the fifth and sixth columns by the tenderers
- If the Contractor/Consultant is to be paid an amount for the item which is not adjusted if the quantity of work changes, the tenderer enters the amount in the Price column only, the Unit, Quantity and Rate columns being left blank.

This is then a lump sum contract

- If the Contractor/Consultant is to be paid an amount for the item of work which is the rate for the work multiplied by the quantity completed, the tenderer enters the rate which is then multiplied by the expected quantity to produce the Price, which is also entered.
- This is then a remeasurement contract.

Target contracts and cost reimbursable contracts are not provided for as they are regarded as unsuitable and too complex for this type of work. Similarly, Management Contracts will not be used on the type of work that the Engineering and Construction Short Contract is designed for.

When considering payment to the Contractor using the Price List, one must consider two terms:

(i) The Prices

These are the various elements that make up the total Price and are the amounts stated in the Price column of the Price List.

(ii) The Price for Work Done to Date

This term is used in making the assessment of amounts due to the Contractor/Consultant, and is the total of:

- the Price for each lump sum item in the Price List which the Contractor/Consultant has completed and
- where a quantity is stated for an item in the Price List, an amount calculated by multiplying the quantity which the Contractor/Consultant has completed by the rate

plus

- other amounts to be paid to the Contractor/Consultant (including any tax which the law requires the Client to pay to the Contractor/Consultant)
- This will include VAT

less

- amounts to be paid by or retained from the Contractor.

For a compensation event that only affects the quantities of work shown in the Price List, the changes to the Prices is assessed by multiplying the changed quantities of work by the appropriate rates in the Price List.

Question 0.34 Can you explain what a Task, and a Task Order are under the NEC4 Term Service Contract?

In order to consider the use of Tasks and Task Orders within the NEC Term Service Contract, we must first consider the various related definitions:

11.2(18) *A Task is work included in the service which the Service Manager instructs the Contractor to carry out and for which a Task Order programme is required.*

11.2(19) *Task Completion is when the Contractor has done all the work in the Task and corrected Defects which would have prevented the Client or Others from using the Affected Property or Others from doing their work.*

11.2(20) *Task Completion Date is the date for completion stated in the Task Order unless later changed in accordance with the Contract.*

11.2(21) *A Task Order is the Service Manager's instruction to carry out a Task.*

The Service Manager may issue a Task Order to the Contractor. Prior to issuing the Task Order the Service Manager instructs the Contractor to submit a quotation for the Task. The instruction includes:

- a detailed description of the work in the Task
- the Task Starting Date and Task Completion date and
- the amount of delay damages for late completion of the Task.

The Contractor is required to submit the quotation within three weeks of being instructed. The Service Manager replies to the quotation within two weeks of receiving it.

The assessment of the Task is in the form of a Task price list. Work covered by rates in the Price List is priced using those rates; other work is priced in the same way as a compensation event is priced.

The reply is:

- acceptance of the quotation and the issue of the Task Order
- an instruction to submit a revised quotation
- that the Service Manager will be making the assessment or
- a notification that the Task will not be instructed.

The Service Manager and the Contractor may mutually agree to extend the period for the Contractor's submission and the Service Manager's reply.

If a Task Order is issued, the Task price list is inserted in the Price List, and the work involved in the Task Order is added to the Scope.

The Task Order programme

Under Clause 33.1, the Contractor is required to submit a Task Order programme to the Service Manager for acceptance within the period stated in the Contract Data.

The Contractor shows on each Task Order programme that he submits for acceptance (see above for similar contents within the submission of the Contractor's plan):

- *the Task starting date and the Task Completion Date*
- *planned Task completion and*
- *the order and timing of the work of the operations which the Contractor plans to do in order to complete the Task*
- *provisions for*
  - *float*
  - *time risk allowances*
  - *health and safety requirements and*
  - *the procedures set out in the contract*

- *the dates when, in order to Provide the Service in accordance with the Task Order, the Contractor will need*
  - *access to the Affected Property*
  - *acceptances*
  - *Plant and Materials and other things to be provided by the Client and*
  - *information from Others*
- *for each operation, a statement of how the Contractor plans to do the work identifying the principal Equipment and other resources which will be used and*
- *other information which the Scope requires the Contractor to show on a Task Order programme submitted for acceptance.*

Note: Quotations for compensation events comprise proposed changes to the Prices and any delay to a Task Completion Date assessed by the Contractor.

Appendix I

Main and Secondary Options within the NEC4 Contracts

ECC – Engineering and Construction Contract
ECS – Engineering and Construction Subcontract
PSC – Professional Service Contract
PSS – Professional Service Subcontract
TSC – Term Service Contract
DBO – Design, Build, Operate Contracts
AC – Alliance Contract
SC – Supply Contract

Main Options (or nearest equivalent wording)

		ECC	ECS	PSC	PSS	TSC	DBO	AC	SC
Option A	Priced contract with activity schedule	✓	✓	✓	✓	✓			
Option B	Priced contract with bill of quantities	✓	✓						
Option C	Target contract with activity schedule	✓	✓	✓	✓	✓			
Option D	Target contract with bill of quantities	✓	✓						
Option E	Cost reimbursable contract	✓	✓	✓	✓	✓			
Option F	Management contract	✓							

Secondary Options (or nearest equivalent wording)

		ECC	ECS	PSC	PSS	TSC	DBO	AC	SC
Option X1	Price adjustment for inflation	✓	✓	✓	✓	✓			✓
Option X2	Changes in the law	✓	✓	✓	✓	✓			✓
Option X3	Multiple currencies	✓	✓	✓	✓	✓	✓		✓
Option X4	Ultimate holding company guarantee	✓	✓	✓	✓	✓	✓	✓	✓
Option X5	Sectional Completion	✓	✓	✓	✓				

Option X6	Bonus for early Completion	✓	✓	✓			✓
Option X7	Delay damages	✓	✓	✓			✓
Option X8	Undertakings to the Client or Others	✓	✓	✓	✓		✓
Option X9	Transfer of rights	✓	✓	✓	✓		✓
Option X10	Information modelling	✓	✓	✓	✓	✓	✓
Option X11	Termination by the Client	✓	✓	✓	✓	✓	✓
Option X12	Multi-party collaboration	✓	✓	✓	✓		✓
Option X13	Performance bond	✓	✓	✓	✓		✓
Option X14	Advanced payment to the Contractor	✓	✓	✓	✓ ✓		✓
Option X15	The Contractor's design	✓	✓	✓	✓ ✓		✓
Option X16	Retention	✓	✓				✓
Option X17	Low performance damage	✓	✓	✓	✓		✓
Option X18	Limitation of liability	✓	✓	✓	✓		✓
Option X19	Termination by either Party	✓		✓	✓	✓	
Option X20	Key Performance Indicators	✓	✓	✓	✓	✓	
Option X21	Whole life cost	✓	✓		✓		
Option X22	Early Contractor involvement	✓			✓	✓	
Option X23	Extending the Service Period			✓	✓		
Option X24	The accounting periods			✓		✓	
Option X25	Supplier warranties			✓			✓
Option X26	Programme of work	✓	✓	✓	✓	✓	✓
Option Y(UK)1	Project Bank Account	✓	✓	✓	✓	✓	✓
Option Y(UK)2	The Housing Grants, Construction & Regeneration Act 1996	✓	✓	✓	✓	✓	✓
Option Y(UK)3	The Contracts (Rights of Third Parties) Act 1999	✓	✓	✓	✓	✓	✓
Option Z	Additional conditions of contract	✓	✓	✓	✓	✓	✓

Chapter 1
Early warnings and Risk Registers

Question 1.1 Is there a standard format within the NEC4 contracts for an early warning notice? Is there any remedy if the Project Manager or the Contractor fails to give an early warning?

First, in direct response to the question, there is no standard format within the NEC4 contracts for early warnings.

Let us consider the NEC4 Engineering and Construction Contract, where early warnings are covered within the ECC by Clauses 15.1 to 15.4.

Early warnings are a key component of the overall risk management process in the ECC. The process is not about liability, but instead about the Parties collaborating to identify, mitigate or remove the effect of matters that could cause difficulty.

The ECC obliges the Project Manager and the Contractor to notify each other as soon as either becomes aware of any matter which could affect the project in terms of time, cost or quality.

The requirement is to *notify*, and this must be done in a form that can be read, copied and recorded, and separately from other communications, in accordance with Clauses 13.1 and 13.7. The obligation is to notify as soon as either becomes aware of a matter, and this can often be difficult for parties to demonstrate one way or the other (see Figure 1.1).

Sometimes correspondence or records may show when the Contractor or Project Manager first became aware of something, but this can be a matter of subjective interpretation. Note that the contract says *"becomes aware"* not *"should have become aware"*.

Under Clause 15.1 the Contractor and the Project Manager give an early warning by notifying the other as soon as either becomes aware of any matter that could:

- *increase the total of the Prices*
   The price of the works, in the form of the activity schedule, the bill of quantities, or the target.
- *delay Completion*
   The completion of the whole of the works.

Contract:	EARLY WARNING NOTICE
Contract No:	EWN No..

Section A: Notice

To : Project Manager/Contractor

Description

This matter could:

- ☐ Increase the total of the Prices
- ☐ Delay Completion
- ☐ Delay meeting a Key Date
- ☐ Impair the performance of the works in use

Early warning meeting called?   Yes/No   Date: ....

Signed:................................(Contractor/Project Manager)   Date: ....

Action by: **Date required:**

Section B: Reply

To: Contractor/Project Manager

Signed:................................(Contractor/Project Manager)   Date: ....

Copied to:

Contractor ☐   Project Manager ☐   Supervisor ☐   File ☐   Other .... ☐

Figure 1.1 Suggested template for early warning notice

- *delay meeting a Key Date*

    The completion of an intermediate "milestone date" in accordance with Clause 11.2(11).
- *impair the performance of the works in use*

    This sometimes causes confusion, but if we take as an example the Contractor is instructed by the Project Manager to use a particular type of pump and the Contractor knows from experience that that pump would probably not be sufficient to meet the Client's requirements once the works are taken over, then the Contractor should give early warning.

The Project Manager or the Contractor may also give an early warning by notifying of any other matter which could increase the Contractor's total cost. One could query whether a matter that could increase the Contractor's cost, but not affect the Prices, should be an early warning matter, or for that matter, whether it should be anything to do with the Project Manager, particularly if Option A or B has been selected.

However, the words are *"the Contractor may give an early warning"* so it is not obliged to do so. This provision is designed to encourage collaboration between the Parties, irrespective of their contractual liability.

Note, also within Clause 15.1, the Contractor is not required to give an early warning for which a compensation event has previously been notified; so, as an example, if the Project Manager gives an instruction that changes the Scope, it is a compensation event, for which neither the Project Manager nor the Contractor are required to give early warning.

The author has encountered situations where Clients and Project Managers appear hostile to the receipt of early warnings from Contractors. Sometimes they are viewed as the first stage in a compensation event process or similar. That may be so, but not always, and in effect could prevent a compensation event occurring or at least lessen its effect.

Statistically speaking one would expect an equal number of early warnings to be issued by the Project Manager and the Contractor. Typically, though, the ones issued by the Contractor frequently outnumber the ones issued by the Project Manager.

Notifying early warnings

The contract requires (Clause 13.1) that all communications, for example instructions, notifications, submissions, etc. are in a form that can be read, copied and recorded, so early warnings should not be a verbal communication such as a telephone conversation. If the first notification is a telephone conversation, or a comment in a site meeting, it should be immediately confirmed in writing to give it contractual significance.

Also, Clause 13.7 requires that notifications that the contract requires must be communicated separately from other communications; therefore, early warnings

Early warnings and Risk Registers 91

must not be included within a long letter that covers a number of issues, or embodied within the minutes of a progress meeting.

There are some key words within the obligation to notify:

- *"The Contractor and the Project Manager"* – No-one else has the authority or obligation to give an early warning. The Project Manager is therefore notifying on behalf of itself, the Client, the Supervisor, the Client's Designers and many possible others who it represents within the contract. The Contractor is notifying on behalf of itself, its Subcontractors, its Designers (if appropriate), and again many possible others who it represents under the contract. Early warnings should be notified by the key people named in Contract Data Part 2.

    Project Managers are often criticised for seeing early warnings as something the Contractor has to do, and in fact most early warnings are actually issued by the Contractor. However, the Contractor and the Project Manager are obliged to give early warnings each to the other, so it is critical that Project Managers play their part in the process.

    As an example, if the Project Manager becomes aware that it will be late in delivering some design information to the Contractor, it should issue the early warning as soon as it becomes aware that the information will not be delivered to the Contractor, not wait and subsequently blame the Contractor for not giving an early warning stating that it has not received the information!

- *"As soon as"* – means immediately. There are a number of clauses within the contract that deal with the situation where the Contractor did not give an early warning. Whilst the party who gives the early warning must do so as soon as it becomes aware of the potential risk, the other Party should respond as soon as possible and in all cases within the period for reply in Contract Data Part 1.

- *"Could"* – not must, will or shall. Clearly there is an obligation to notify even if it is only felt something *may* affect the contract, but there is no clear evidence that it will.

The Project Manager enters early warning matters in the Early Warning Register. If the Project Manager gives an instruction for which a compensation event has already been notified, there is no requirement for either party to give an early warning.

It must be emphasised that early warnings are not the first step toward a compensation event as is often believed. Early warnings feature in a completely separate section of the contract and in fact the early warning provision is intended to prevent a compensation event occurring or at least to lessen its effect. It can also be used to notify a problem that is totally the risk of the notifier. It is also worth mentioning that early warnings are a notice of a future risk, not a past one. The parties are not required, nor is it of any value, to notify a risk that has already happened.

Question 1.2   What is an Early Warning Register, and what is its purpose within the NEC4 Engineering and Construction Contract?

The NEC4 contracts provide for the use of a risk register, now referring to it as an "Early Warning Register", to distinguish it from a Project Risk Register which would take into account other risks such as health, and safety, environment, etc.

The Contract Data contains matching sections in Parts 1 *and* 2 for the Parties to add matters that will be included in the Early Warning Register. That is to say that both Parties have the ability to list risks that they wish to form part of the risk management processes in the contract. These **do not** change the risk allocation between the Parties. These entries are to assist the risk management processes of the Parties by listing those risks that collaborative behaviour will help with.

It is critical that the Parties fully understand that the purpose of an Early Warning Register is to list all the identified risks and the results of their analysis and evaluation. It can then be used to track, review and monitor risks as they arise to enable the successful completion of the project. The Early Warning Register does not allocate risk, that is done by the contract.

The Early Warning Register is a simple document, but a vital one in the process of risk management. Its role and contents are frequently misunderstood by Parties, particularly where people confuse its use with risk registers elsewhere, such as those that are components of company management systems. Its purpose is to assist the Parties and the Project Manager with managing risk in the project delivery and not to allocate responsibility or blame.

It is helpful to start with its definition from clause 11.2(8):

The Early Warning Register is a register of matters which are

- *listed in the Contract Data for inclusion and*
- *notified by the Project Manager or the Contractor as early warning matters.*

It includes a description of the matter and the way in which the effects of the matter are to be avoided or reduced.

The Early Warning Register contains information about these risks:

- Those listed in the Contract Data
    - by the Client in Contract Data Part 1
    - by the Contractor in Contract Data Part 2
- Those notified as an early warning matter
    - by the Project Manager
    - by the Contractor

The information contained in the Early Warning Register is:

- a description of the risk
- the actions to be taken to avoid or reduce the risk.

The Early Warning Register need not contain anything else. It is a document produced after the Contract Date and therefore it does not form part of the contract.

For practical reasons though it should not contain any statements that purport to allocate risk to one Party or the other, that will just prove confusing in the long run.

The contract does not prescribe the format or layout of the Early Warning Register, other than to list the risks in the contract and those that come to light at a later date and are notified as early warnings following which, if there is an early warning meeting, the Project Manager revises the Early Warning Register to record the outcome of the meeting.

Note that by the definition within Clause 11.2(8), risks that were not originally included in the Contract Data or subsequently notified as early warnings should not be included in the Early Warning Register. Contract Data Part 2 allows the Contractor to identify matters that will be included in the Early Warning Register. The Early Warning Register does not allocate or change the risks in the contract, it records them and assists the Parties in managing them. In that sense it is a valuable addition to the workings of the contract.

In order to compile the Early Warning Register the risks must first be listed, then they are quantified in terms of their likelihood of occurrence and their potential impact upon the project.

In that sense, the contract does not prescribe the format or layout of the Early Warning Register, or its intended purpose, other than to list the risks in the contract and those that come to light at a later date and are notified as early warnings, following which, if there is an early warning meeting the Project Manager revises the Early Warning Register to record the outcome of the meeting.

In effect, the Contractor could seek to qualify its tender by including additional matters within Contract Data Part 2 at the time of tender, which it sees, or would like to see, as Client's liabilities. That is not the intent of this provision, but Clients should check for such entries during tender appraisal and whether there is a potential impact upon the Contractor's tender and/or the project.

As stated previously the Early Warning Register does not allocate or change risks in the contract.

Figure 1.2 shows an extract from a typical register which would comply with the contract in that it includes the basic requirement for a description of the risk and a description of the actions that are to be taken to avoid or reduce the risk.

EARLY WARNING REGISTER

Contract:..................
Contract No:..............
Contractor:................
Project Manager:........

Description of risk	Implications	Likelihood of occurrence (1–5) (Least – Most)	Potential impact (1–5) (Low – High)	Risk score	Risk owner	Mitigation strategy By whom? By when?	Allowance in the total of the prices	Programme allowance	Employer cost allowance	Risk status	Last updated
Discovery of unforeseen existing silo bases	Delay and additional costs in removal	2	4	8	Initially Contractor, unless additional silo bases discovered, then Employer	Site Information identifies likely locations Contractor to take due regard	No allowance for unforeseen	No allowance for unforeseen	£20,000	Reducing Excavation 50% complete No additional silo bases discovered to date	15/12/2018

Figure 1.2 Typical Early Warning Register

There is no stated list of components of an Early Warning Register, but column headings should typically be titled:

1   Description of risk

A clear description of the nature of the risk, if necessary referring to other documents such as site investigations, etc.

2   Implications

What would happen if the risk were to occur?

3   Likelihood of occurrence

This provides an assessment of how likely the risk is to occur. The example shows a forecast on a 1 (least likely) to 5 (most likely) basis, though it may be assessed as percentages, colour coding, or simply "Low" (less than 30% likelihood), "Medium" (31–70% likelihood), "High" (more than 70% likelihood).

4   Potential impact

This assesses the impact that the occurrence of this risk would have on the project in terms of time and/or cost. The example shows the assessment on a 1 (low) to 5 (high) basis.

5   Risk score

Risk = Likelihood of something occurring × Impact should it occur.

By evaluating risk on that basis each can be evaluated and categorised into an order of importance. This formula can only be applied to economic loss, and different standards would need to be adopted when considering risk of death or serious bodily injury.

6   Risk owner

This identifies which Party or individual owns the risk; this is allocated through the contract, the principle being that if the Contractor owns it, it is deemed to have allowed for its price and/or time effect within its tender. If the Client owns it, it will be a compensation event.

7   Mitigation strategy

This registers what actions are proposed which could be taken to prevent, reduce, or transfer the risk. Also, who is responsible for this action, and when should the action take place?

8 Allowance in the total of the Prices

How much has the Contractor allowed in its tender for a risk it owns?

9 Programme allowance

How much time has the Contractor allowed in its programme for a risk it owns?

10 Client cost allowance

If it is a Client's risk, how much has it allocated to cover a potential compensation event?

11 Risk status

This identifies whether the risk is current, and also whether it is increasing, decreasing, or has not changed since it was last reviewed?

12 Last updated

When was an additional risk identified? When was the Early Warning Register last updated?

Question 1.3 What is the purpose of an early warning meeting within the NEC4 contracts?

In previous versions of NEC prior to NEC3, the meeting was termed an "early warning meeting"; then the NEC3 called it a "risk reduction meeting", although many practitioners stated that "early warning meeting" would be the better term as it is tied into the early warning process. Now NEC4 refers to it again as an "early warning meeting".

The Early Warning Register becomes more important once early warnings lead to early warning meetings; see Clauses 15.2, 15.3 and 15.4.

The Project Manager prepares a first Early Warning Register and issues it to the Contractor within one week of the starting date.

The Project Manager also instructs the Contractor to attend a first early warning meeting within two weeks of the starting date, further early warning meetings at no longer than the durations stated in the Contract Data, until Completion of the whole of the works.

Clause 15.2 includes a rare example of an optional action in an NEC contract, when, after the first early warning meeting, either the Project Manager or the Contractor *may* instruct the other to attend an early warning meeting.

Note the key word in Clause 15.2 is "instruct". If the Contractor calls an early warning meeting, it is *instructing* not merely requesting the Project Manager to attend. This clause is clearly intended to promote ownership of the project by the Contractor and the Project Manager, and any issues that could affect it, together with their resolution. Hopefully in the real world such meetings are arranged by co-operation and invitation.

Notwithstanding that the Project Manager or the Contractor can instruct the other to attend an early warning meeting, early warning meetings are also held on a periodic basis no longer than the interval stated in the Contract Data until Completion of the whole of the works.

An early warning meeting may only have the Project Manager and the Contractor present, though each may instruct other people to attend if the other agrees. Also, a Subcontractor may attend, so in reality a number of people normally attend the meeting.

On a lighter note, the "rule of meetings" will often apply in that the productivity of the meeting is often inversely proportional to the number of people who attend it! However, this should not mask the importance of getting the right people to attend. Project delivery, particularly in buildings, relies on multiple specialists and these should be encouraged to participate wherever necessary.

Clause 15.3 describes what the attendees at an early warning meeting should do in an effort to either resolve the problem or at least attempt to resolve it:

- making and considering proposals for how the effect of each matter in the Early Warning Register can be avoided or reduced,
- seeking solutions that will bring advantage to all those who will be affected,
- deciding on the actions which will be taken and who, in accordance with this contract will take them,
- deciding which matters can be removed from the Early Warning Register and
- reviewing actions recorded in the Early Warning Register and deciding if different actions need to be taken and who, in accordance with the contract, will take them.

Clearly the purpose of the early warning meeting is to actively consider ways to avoid or reduce the effect of the matter that has been notified. In some cases, the matter can be fully resolved, but in others, as the matter may not yet have occurred it may simply have to be "parked" and recorded as such in the Early Warning Register. It is the Project Manager's responsibility to revise the Early Warning Register to record the outcome of the meeting, and to issue the revised Early Warning Register to the Contractor.

Early warning meetings can happen in several ways;

- As a standing agenda item in regular (say weekly, or monthly) meetings
- As a routine meeting (say once a month)
- Ad hoc, when the matter is sufficiently urgent.

98 Early warnings and Risk Registers

Meetings can provide a distraction to the "day job", so complying with Clause 15.2 should not place unnecessary demands on the participants. The use of telephone, video or online conferencing works well, particularly with multiple participants in different locations.

Clause 15.3 describes the agenda for an early warning meeting and it is a relatively predictable risk management process that leads to the Project Manager and Contractor agreeing actions to remove or mitigate a risk.

As stated above, the contract requires those attending to cooperate in:

- Making and considering proposals
- Seeking solutions
- Deciding on actions
- Deciding which risks have been avoided or passed and can therefore be removed from the Early Warning Register.

This final bullet point is the conclusion of the process for individual matters. Clause 15.4 goes on to say that the Project Manager revises the Early Warning Register after the early warning meeting and issues a copy to the Contractor within one week of the meeting. Whilst clause 15.3 mentions removing risks from the Early Warning Register the author recommends retaining them, perhaps in an appendix or ~~struck through~~, for future consultation.

Chapter 2

Contractor's design

Submitting design proposals, liability for design, etc.

Question 2.1   We wish to use the NEC4 Engineering and Construction Contract and the NEC4 Engineering and Construction Short Contract for a number of design and build projects. Can we do this and if so, how?

The NEC contracts, and in this case specifically the NEC4 Engineering and Construction Contract and the NEC4 Engineering and Construction Short Contract are unique amongst construction contract families in that, whilst most other contracts provide for portions, or all of the design, to be carried out by the Contractor, they all use a separate contract for design and build where the Contractor has full responsibility for design, whereas the NEC4 contracts use the same contract whether design is to be carried out by the Client, the Contractor or a combination of the two, as responsibility for design, and any associated design criteria, performance requirements and obligations are defined within the Scope.

In reality, there is no need for a separate design and build contract as within the NEC contracts the clauses that cover early warnings, programme, payment, change management, etc. are the same whether or not the Contractor has design responsibility (see Figure 2.1).

If we consider the NEC4 Engineering and Construction Contract, the "default position" within the contract is that all design is carried out by the Client, but any design to be carried out by the Contractor is assigned to him within the Scope, and Clauses 21 to 23 cover design obligations, and submission and acceptance of the Contractor's design proposals.

In the Engineering and Construction Contract "the Contractor Provides the Works in accordance with the Scope" (Clause 20.1) and "the Contractor designs the parts of the works which the Scope states the Contractor is to design" (Clause 21.1).

Any design to be carried out by the Contractor within the Engineering and Construction Short Contract is also assigned within the Scope in the same way.

It is clear, then, that the Scope should clearly show what, if anything, is to be designed by the Contractor and this could consist of detailed specifications or some form of performance criteria and certain warranties.

100 Contractor's design

When preparing the Scope, which includes Contractor's design, it is important that the right balance of information should be considered. If it is too prescriptive it will lead to all the tendering Contractors submitting similar designs and prices. If the Scope is not detailed enough, then none of the tendering Contractors will produce a design that is as the Client intended.

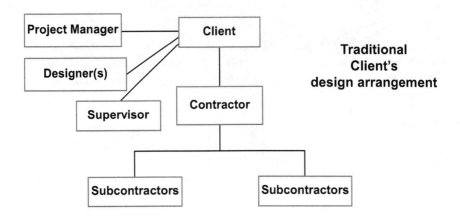

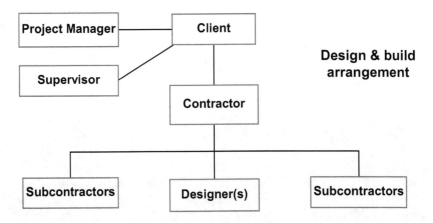

Figure 2.1 Design and Build using the NEC4 Engineering and Construction Contract

Question 2.2   In a design and build contract using the NEC4 Engineering and Construction Contract, which takes precedence, the Client's Requirements or the Contractor's Proposals?

The NEC4 contracts are unlike most other construction contracts in that they do not provide a priority or hierarchy of documents, but one can "read into" the various clauses to determine which takes precedence in the event of an inconsistency.

A question to then be considered is, if the Contractor has full design responsibility and there is an inconsistency between the Scope provided by the Client (normally referred to in other contracts as the "Employer's Requirements") and the Scope provided by the Contractor (normally referred to in other contracts as the "Contractor's Proposals"), which takes precedence?

The answer can be found in Clauses 11.2(6) and 60.1(1).

Defects are defined in Clause 11.2(6) as:

- *a part of the works which is not in accordance with the Scope or*
- *a part of the works designed by the Contractor which is not in accordance with the applicable law or the Contractor's design which the Project Manager has accepted.*

From bullet point 1 above, if the Contractor's design does not comply with the Scope, then it is a Defect, the Scope clearly taking precedence over the Contractor's design.

From bullet point 2 above, the Contractor's design must also comply with the law and anything that the Project Manager has already accepted.

Under Clause 60.1(1), the Project Manager gives an instruction changing the Scope except

- *a change made in order to accept a Defect or*
- *a change to the Scope provided by the Contractor for his design which is made*
    - *at the Contractor's request or*
    - *in order to comply with other Scope provided by the Client.*

From the above bullet point 2, if the Contractor has to change his design to comply with the Scope provided by the Client, that is not a compensation event, again reinforcing the fact that the Scope takes precedence.

Question 2.3   If the Project Manager on an NEC4 Engineering and Construction Contract accepts the Contractor's design and it is later found that the design will not work, or it is not approved by a third party regulator, who is liable?

Under Clause 21.2, the Contractor is required to submit the particulars of his design as the Scope requires to the Project Manager for acceptance.

The particulars of the design in terms of drawings, specification and other details must clearly be sufficient for the Project Manager to make the decision as to whether the particulars comply with the Scope and also, if relevant, the applicable law. The Scope may stipulate whether the design may be submitted in parts, and also how long the Project Manager requires to accept the design.

Note that in the absence of a stated time scale for acceptance/non acceptance of design, the "period for reply" will apply.

Many Project Managers are concerned that they have to "approve" the Contractor's design and therefore they are apprehensive as to whether they would be qualified, experienced, and also insured, to be able to do so. Note the use of the word "acceptance" as distinct from "approval". Acceptance denotes compliance with the Scope or the applicable law; it does not denote that the design will work, that it will be approved by regulating authorities, or that it will fulfil all the obligations that the contract and the law impose. Therefore performance requirements such as structural strength, insulation qualities, etc. do not need to be considered.

A reason for the Project Manager not accepting the Contractor's design is that it does not comply with the Scope or the applicable law. The Contractor cannot proceed with the relevant work until the Project Manager has accepted the design.

Note that under Clause 14.1, the Project Manager's acceptance does not change the Contractor's responsibility to Provide the Works or his liability for his design. This is an important aspect as, if the Project Manager accepts the Contractor's design but following acceptance there are problems with the proposed design, for example in meeting the appropriate legislation or the requirements of external regulatory bodies, this is the Contractor's liability.

Note that, under Clause 27.1, the Contractor is required to obtain approval of his design from Others where necessary. This will include planning authorities, and other third party regulatory bodies.

Note also, Clause 60.1(1) second bullet point, which clearly refers to the fact that the Client's Scope prevails over the Contractor's Scope. This clause gives precedence to the Scope in Part 1 of the Contract Data over the Scope in Part 2 of the Contract Data. Thus, the Contractor should ensure that the Scope he prepares and submits with his tender as Part 2 of the Contract Data complies with the requirements of the Scope in Part 1 of the Contract Data.

Question 2.4   We wish to novate a design consultant from the Client to the Contractor under an NEC4 Engineering and Construction Contract with full Contractor's design. Can we do this?

Whilst the principle of design and build agreements is that the Client preparers his requirements and sends them to the tendering Contractors and they prepare their proposals to match the Client's requirements, in reality in nearly half of design and build contracts the Client has already appointed a design team that prepares feasibility proposals and initial design proposals before tenders are invited.

Outline planning permission and sometimes detailed permission may have also been obtained for the scheme before the Contractor is appointed. Each Contractor then tenders on the basis that the Client's design team will be novated or transferred to the successful tendering Contractor who will then be responsible for appointing the team and completing the design under a new agreement.

This process is often referred to as "novation", which means "replace" or "substitute" and is a mechanism where one Party transfers all its obligations and benefits under a contract to a third party. The third party effectively replaces the original Party as a Party to that contract, so the Contractor is in the same position as if he had been the Client from the commencement of the original contract.

Many prefer to use the term "consultant switch" where the design consultant "switches" to work for the Contractor under different terms as a more accurate definition. This approach allows the Client and his advisers time to develop their thoughts and requirements, consider planning consent issues, then when the design is fairly well advanced, the designers can be passed to the successful design-and-build Contractor.

The NEC4 contracts do not include pro forma novation agreements, but it is critical that the wording of these agreements be carefully considered as there are many badly drafted agreements in existence.

Novation should be by a signed agreement or by deed. As with all contracts, there must be consideration, which is usually assumed to be the discharge of the original contract and the original Parties' contractual obligations to each other. If the consideration is unclear, or where there is none, the novation agreement should be executed as a deed. In many cases the agreement states briefly (and very badly) that from the date of the execution of the novation agreement, the Contractor will take the place of the Client as he had employed the designers from the beginning, the document stating that in place of the word "Client" one should read "Contractor"; however, the issues of design liability, inspection, guarantees and warranties may not always apply on a back to back basis, so one must take care to draft the agreement in sufficient detail and refer to the correct Parties.

In practice, the best way is to have an agreement drafted between the Client and the Designer covering the pre-novation period, and a totally separate agreement drafted between the Contractor and the Designer for the post-novation period. When a contract is novated, the other (original) contracting party must be left in the same position as he was in prior to the novation being made. Essentially, a novation requires the agreement of all three Parties.

As this book is intended as a guide for NEC4 practitioners worldwide, it is not intended to examine specific legal cases within the UK, as they may not apply on an international basis; but there is certainly case law available in terms of design errors in the pre-novation phase being carried through as the liability of the Contractor in the post-novation stage, and also the accuracy and validity of site investigations and the Client's and the Contractor's obligations in terms of checking and taking ownership of such information.

Question 2.5   The Contractor has designed a temporary access walkway bridging across two areas of the Site. He has not submitted his design proposals for the Project Manager's acceptance before constructing the walkway, but we insist that he is in breach of contract by failing to do so. What can we do to remedy this breach?

If the Contractor has designed an item of Equipment, for example a temporary access way, or specialist scaffold, the Project Manager, under Clause 23.1 *may* instruct him to submit particulars to him for acceptance, so the submission is not obligatory as with the design of parts of the works.

So, in this case, unless the Project Manager gave that instruction, the Contractor is not in breach of contract by failing to submit particulars of his design.

Notwithstanding this, the Contractor's design, whether the Project Manager has given the instruction or not, the piece of work still has to be built so that it is safe and fully compliant with any relevant legislation.

The Project Manager may not accept the design if it does not comply with the Scope, the Contractor's design which the Project Manager has accepted, or the applicable law.

Chapter 3

Time and the Accepted Programme

The submission or non-submission of a programme, float and time risk allowances, method statements, etc.

Question 3.1 Can we include liquidated or unliquidated damages in an NEC4 contract? How are these deducted in the event of delayed Completion?

Liquidated damages in other construction contracts are referred to as delay damages in the NEC4 contracts, and in the NEC4 Engineering and Construction Contract are covered by Secondary Option X7.

Delay damages are predefined amounts inserted into the contract and paid or withheld from the Contractor in the event that he fails to complete the works by the Completion Date. The amount included within the contract for delay damages should be a "genuine pre estimate of likely loss", i.e. should not constitute a penalty, although other legislations outside the UK permit penalties.

Many non NEC4 contracts require the Contract Administrator (Engineer, Architect, etc.) to issue some form of certificate confirming that the Contractor failed to complete on time, and also for the Client to notify the Contractor in writing that he will be withholding the relevant damages, but the NEC4 Engineering and Construction Contract provides for the Contractor to pay delay damages at the rate stated in the Contract Data until Completion or the date on which the Client has taken over the works, whichever is earlier, without having to certify that the Contractor has defaulted.

Also, many contracts require the Contract Administrator to value the works and advise the Client of its entitlement to liquidated damages, and the Client then deducts the liquidated damages from the amount due to the Contractor. The NEC4 Engineering and Construction Contract, however, requires the Project Manager to deduct amounts to be paid or retained from the Contractor in respect of delay damages within his assessment and certificate.

Under Clause X7.3, if the Client takes over the works before Completion, the Project Manager assesses the benefit to the Client of taking over that part of the works as a proportion of taking over the whole of the works and the delay damages are reduced in this proportion.

Whilst it is probably the most correct way to assess remaining delay damages, the Project Manager having to assess the benefit to the Client can at best be a subjective exercise, and may possibly lead to disputes with the Contractor.

Most other non NEC4 contracts state that the amount of liquidated damages is reduced by the same proportion the part that has been taken over bears to the value of the whole of the works, so if a third of the value of the works has been taken over, the amount of the liquidated damages is reduced by a third.

If Secondary Option X7 is not selected, and the Contractor fails to complete the works by the Completion Date, it may still be liable to the Client for unliquidated damages, though in this case the Client would have to prove and quantify any loss or cost incurred as a result of that failure, and also of course try to recover those damages from the Contractor as the contract does not allow the Project Manager to deduct amounts to be paid or retained from the Contractor or for the Client to withhold payment in this respect.

Question 3.2   Do the NEC4 contracts have provision for Sectional Completion where the Client may wish to take over parts of the works before completion of the whole project?

If the Client requires sections of the works to be completed by the Contractor before the whole of the works are completed, then Secondary Option X5 should be chosen.

References in the contract to the "works", "Completion" and "Completion Date" will then apply to either the whole of the works, or any section of the works.

Contract Data Part 1 should give a description of each section and the date by which it is to be completed. Option X5 may be selected together with Option X6 (bonus for early Completion) and/or Option X7 (delay damages).

Question 3.3   In an NEC4 Engineering and Construction Contract, what is the relationship between "Completion" and "Take over"?

In order to consider the relationship between these two terms, let us first consider their definitions under the Contract:

Completion

The Engineering and Construction Contract defines Completion under Clause 11.2(2) as when the Contractor has:

- *done all the work which the Scope states is to be done by the Completion Date and*
- *corrected notified Defects which would have prevented the Client from using the works or Others from doing their work.*

Take over

The Client need not take over the works before the Completion Date if the Contract Data states it is not willing to do so. Otherwise, the Client takes over the works not later than two weeks after Completion.

The Client may use any part of the works before Completion has been certified. If he does so, he takes over the part of the works when he begins to use it except if the use is

- *for a reason stated in the Scope or*
- *to suit the Contractor's method of working.*

The Project Manager certifies the date upon which the Client takes over any part of the works and its extent within one week of the date.

Once the Contractor has completed the works, the Client takes over within two weeks. This occurs even if the Contractor completes the works early. However, there is an optional statement within Contract Data Part 1 that "the Client is not willing to take over the works before the Completion Date". If this is selected then if the Contractor completes the works early he still has responsibility for the works until the Completion Date.

Note that a period is stated in Contract Data Part 1 as the period between Completion of the whole of the works and the Defects date.

If Option X7 is selected, the liability for delay damages is at the earlier of

- *Completion and*
- *the date on which the Client takes over the works.*

Partial possession

Whilst many contracts have express provision for partial possession by the Client, with the consent of the Contractor which shall not be unreasonably withheld, the Engineering and Construction Contract does not, though under Clause 35.2 the Client may use any part of the works before Completion has been certified.

Note that, if the Project Manager certifies take over of a part of the works before Completion and the Completion Date, then it is a compensation event under Clause 60.1(15). If the Client takes over the works before Completion but after the Completion Date then the Contractor is already late in completing and therefore that would not be a compensation event.

There is no provision for the Contractor to give or refuse consent to the Client taking over the works. The Project Manager certifies the date of take over, or partial take over within one week of it taking place.

Question 3.4   We have a project where we require the Contractor to provide "as built" drawings, maintenance manuals and staff training

as part of the contract. How do we include this requirement within an NEC4 Engineering and Construction Contract?

Once the project is built, it is often very difficult for Clients to obtain supporting documents from the Contractor, such as "as built" drawings, maintenance manuals, software licences, etc.

In order to consider and remedy this issue, let us first consider how the NEC4 contracts consider "Completion".

Whilst many contracts use the terms "Practical Completion" and "Substantial Completion", the NEC4 contracts do not, as these terms are often subject to various interpretations, including whether the Client can take beneficial occupation of the "completed" works.

The NEC4 Engineering and Construction Contract, for example, defines Completion under Clause 11.2(2) as when the Contractor has:

- done all the work which the Scope states is to be done by the Completion Date; and
- corrected notified Defects which would have prevented the Client from using the works and Others from doing their work.

The first bullet of Clause 11.2(2) requires the Contractor to comply with the requirements of the Scope and supersedes traditional terms such as "practical completion". This requirement may include not only physical completion of the Works, but also submission of "as built" drawings, maintenance manuals, software licences, training requirements, successful testing requirements, etc., so it is important to list those items within the Scope.

By including these requirements within the Scope, Completion has not been achieved until they are completed. If Option X7 (Delay Damages) has been selected, the Contractor would also be liable for delay damages until the earlier of

- Completion and
- the date on which the Client takes over the works.

Referring to the second bullet of Clause 11.2(2), the works may contain Defects, though these are not significant enough to prevent the Client from practically and safely using the works, which are fit for the intended purpose. Completion is certified with the requirement that the Contractor corrects the Defect before the end of the Defect correction period following Completion.

There is also a fall back within Clause 11.2(2) in that if the work which the Contractor is to do by the Completion Date is not stated in the Scope, Completion is when the Contractor has done all the work necessary for the Client to use the works and for Others to do their work. This broadly aligns with the traditional terms "Practical Completion" and "Substantial Completion", where the Client is able to take beneficial occupation of the works and use them as intended.

The Project Manager is responsible for certifying Completion, as defined in Clause 11.2(2), within one week of Completion. Normally, the Contractor will request the certificate as soon as he considers he is entitled to it, but such a request is not essential.

Question 3.5   We have a project under the NEC4 Engineering and Construction Contract Option C (target contract with activity schedule) to build a new college where we will require the Contractor to complete wall finishings to classrooms in order that a third party directly employed by the Client will be able to install audio visual equipment fixed to the walls. How can we incorporate this requirement into the contract?

First, let us just pick up on some terminology within the NEC4 contracts.

"Equipment" is defined under Clause 11.2(9) as "items provided and used by the Contractor to Provide the Works and which the Scope does not require the Contractor to include in the works".

So, in NEC terminology, the audio visual equipment should be referred to as either Plant or Materials which are defined under Clause 11.2(14) as "items intended to be included in the works"?

The requirement for the Contractor to complete certain works can be included within an NEC4 Engineering and Construction Contract by using Key Dates.

Within the Engineering and Construction Contract this is covered by Clause 11.2(11), a Key Date being defined as "*the date by which the work has to meet the Condition stated*".

It is an optional clause allowing the Client, should he wish, to include dates when the Contractor must complete certain items of work, possibly to allow others to carry out other work. If used, the Condition and the applicable Key Date are defined by the Client in Contract Data Part 1.

"Key Dates" must be differentiated from "Completion" or "Sectional Completion", in that, with Key Dates, the Client does not take over the works, the works stay under the possession and control of the Contractor, whereas with Completion and Sectional Completion, the Client takes over the works. In this respect, there are no predetermined delay damages applied to the Contractor who does not meet a Condition by a Key Date.

However, if the Project Manager decides that the Contractor has not met the Condition stated by the Key Date, and the Client incurs additional cost on the same project as a result of that failure, then the Contractor will be liable to pay that amount (Clause 25.3). By referring to "the same project", the Client cannot claim costs incurred on another project as a result of the failure of the Contractor to meet a Key Date. This cost is assessed by the Project Manager within four weeks of when the Contractor actually does meet the Condition for the Key Date.

Example

The Contractor is building a new outpatients department for a local hospital. The Client wishes to have new x-ray machinery installed by a specialist which he will employ directly and who will install the machinery as the building work progresses.

In order to do that, the Contractor has to complete the part of the building that will house the machinery and install the necessary electrical power facilities so that the machinery can be tested prior to completion of the project. The necessary work to be carried out by the Contractor in readiness for the specialist to carry out his part will be the Condition and this will have a Key Date attached to it.

If the Contractor then fails to meet the Condition by the Key Date and the Client incurs a cost in having to postpone the installation of the x-ray machinery, then this cost is paid by the Contractor. The cost must be incurred on the same project, so, for example, if the late installation of the x-ray machinery incurs a cost on another project, then this is not paid by the Contractor.

Question 3.6 The Contractor on an NEC4 Engineering and Construction Contract was required to submit a programme with his tender. He has stated that as we accepted his tender, then we have also accepted his programme, and this is therefore the first Accepted Programme. But how can it be so if it does not comply with Clause 31.2?

Clients should instruct Contractors to submit at least outline programmes with their tenders, though it must be appreciated that as the Contractor may not be successful with his tender, it will serve only as an indicative outline of how the Contractor would carry out the works should his tender be successful.

The NEC4 Engineering and Construction Contract provides as an optional statement for the Contractor to submit a first programme for acceptance within a specified period of time after the Contract Date.

So, the two alternatives for the Contractor's submission of his first programme for acceptance are:

1. He may submit, or be required to submit, a programme with his tender in which case it is referenced by the Contractor in Contract Data Part 2.

   Note that under Clause 11.1(1) "the Accepted Programme is the programme identified in the Contract Data or is the latest programme accepted by the Project Manager". Therefore, any programme referred to in

Contract Data Part 2 automatically becomes the first Accepted Programme even though it will probably not comply fully with the requirements of Clause 31.2.
2   If a programme is not identified in Contract Data Part 2, the Contractor submits a first programme to the Project Manager for acceptance within the period of time after the Contract Date, this period of time being stated in Contract Data Part 1.

Question 3.7 Who owns the float in an Accepted Programme within an NEC4 Engineering and Construction Contract?

The Contractor is required, on each programme submitted to the Project Manager for acceptance, to show provisions for float (Clause 31.2).

Float is any spare time within the Contractor's programme, after time risk allowances have been included, and represents the amount of time that operations may be delayed without delaying following operations and/or planned Completion.

It can also represent the time between when the Contractor plans to complete the project and when the contract requires him to complete the project (terminal float).

Float absorbs to a certain extent the Contractor's own delays or the delays caused by a compensation event, thereby lessening or avoiding any delay to planned Completion. In effect, no delay arises unless float on the relevant and critical operations reduces to below zero. Programming is never an exact science, so float gives some flexibility to the Contractor in respect of incorrect forecasts or his own inefficiencies. As the work progresses the float will change as output rates change, Contractor's risk events take place and also compensation events arise.

The general belief, certainly amongst many Contractors, is that float belongs to them, as they wrote the programme, and therefore have the right to work to that programme and use any float that it contains.

Conversely, Clients believe that if it is free time then they have the right to use it, but it actually depends where the float is and what it is for. In principle, float other than terminal float or time risk allowances is a shared resource, it is spare time, it belongs to the project, and therefore may be used by whichever Party needs it first.

There are three primary types of float:

1   the amount of time that an operation can be delayed before it delays the earliest start of following operation (free float);
2   the amount of time that an operation can be delayed before it delays the earliest completion of the works (total float);
3   the amount of time between planned Completion and the Completion Date (terminal float).

Generally, with other non NEC contracts, if the Contractor shows that he plans to complete a project early, then he is prevented from completing as early as he planned, but he still completes before the contract completion date, then although there may be an entitlement to an extension of time under the contract, for example exceptionally adverse weather, none will be awarded as there is no delay to contract completion, but he may have a right to financial recovery subject to him proving loss and/or expense. The Engineering and Construction Contract deals with this in a different way.

If the Contractor shows on his programme planned Completion earlier than the Completion Date and he is prevented from completing by the planned Completion Date by a compensation event, then, when assessing the compensation event, Clause 63.5 states "a delay to the Completion Date is assessed as the length of time that, due to the compensation event, planned Completion is later than planned Completion as shown on the Accepted Programme current at the dividing date", therefore any terminal float is retained by the Contractor, the period of delay being added to the Completion Date to determine the change to the Completion Date.

Any delay to planned Completion due to a compensation event therefore results in the same delay to the Completion Date. Therefore, the extension is granted on the basis of time the Contractor is delayed, i.e. entitlement, not on how long he needs to achieve the current Completion Date.

Clause 63.5 also states "a delay to a Key Date is assessed as the length of time that, due to the compensation event, the planned date when the condition stated for a Key Date will be met is later than the date shown on the Accepted Programme current at the dividing date".

Example

The Completion Date is 30 November 2018, but the Contractor is planning to complete by 16 November 2018, the programme showing the last operation (roof covering) being completed on that date.

The Project Manager has accepted the Contractor's programme. At the beginning of October 2018, the Project Manager gives an instruction to change the specification for the roof from concrete tiles to slates. This is a change to the Scope and therefore a compensation event under Clause 60.1(1).

The Contractor prepares his quotation and finds that he cannot get the slates delivered until week commencing 19 November 2018, therefore the roof cannot be completed until 23 November 2018, and the Contractor then shows in his quotation a delay to planned Completion of one week.

> As Clause 63.5 states that "a delay to the Completion Date is assessed as the length of time that, due to a compensation event, planned Completion is later than planned Completion as shown on the Accepted Programme", assuming the Project Manager accepts the quotation, planned Completion becomes 23 November 2018 and the Completion Date becomes 7 December 2018.

Question 3.8 The Contractor on an NEC4 Engineering and Construction Contract has failed to submit a programme to the Project Manager. What can we do to remedy this? Is there any remedy where the Project Manager fails to respond to a programme submission within the required two weeks?

The ECC has quite comprehensive and specific requirements for a programme, in the form stated in the Scope, to be submitted by the Contractor to the Project Manager, these requirements being defined by nine bullet points within Clause 31.2.

Some practitioners say that the requirement within Clause 31.2 is very onerous and in some cases excessive; however, it is critical that the programme shows in a clear and transparent fashion what the Contractor is planning to do, when and how long each operation will last, and what it needs for the Client and others to be able to comply with it.

There is nothing within the requirements of Clause 31.2 that a competent Contractor would not ordinarily include within a professionally constructed programme submitted for the benefit of itself and the receiving party. Whether and how it chooses to show the information could be another matter!

There are two alternatives for the Contractor's submission of its first programme for acceptance:

1 It may submit, or be required to submit, a programme with its tender in which case it is referenced by the Contractor in Contract Data Part 2. Note that under Clause 11.1(1) "the Accepted Programme is the programme identified in the Contract Data or is the latest programme accepted by the Project Manager". Therefore, any programme referred to in Contract Data Part 2 and submitted by the Contractor as part of its tender, automatically becomes the first Accepted Programme, even though it is unlikely that it will comply with the requirements of Clause 31.2.

2 If a programme is not identified in Contract Data Part 2, the Contractor submits a first programme to the Project Manager for acceptance within the period of time after the Contract Date, this period of time being stated in Contract Data Part 1. This will be a programme intended to comply with Clause 31.2, and is the first "meaningful" programme, even if one is submitted with the Contractor's tender.

Clause 31.2

The *Contractor* shows on each programme which it submits for acceptance

- *the starting date, access dates, Key Dates and Completion Date*

    As stated previously, these dates are stated by the Client in Contract Data Part 1. With regard to Completion Date, if the Client has decided the Completion Date for the whole of the works then it is inserted into Contract Data Part 1. Alternatively, if the Contractor is to decide the Completion Date it inserts this into Contract Data Part 2 as part of its tender.

- *planned Completion*

    This is the date the Contractor *plans* to complete the Works; this is different to the date by which the Contractor is *required* to complete the Works as stated in Contract Data Part 1, or as may be revised in accordance with the contract.

    Although the Contractor can complete any time between its planned Completion Date and the Completion Date, failure to complete on or before the Completion Date constitutes a breach of the Contractor's obligation and if Option X7 has been selected, the Contractor is liable to pay the Client delay damages.

    The difference between the dates when the Contractor *plans* to complete and when it is *required* to complete is float within the programme. This float, often referred to as "terminal" float, belongs to the Contractor and is attached to the whole programme. If a compensation event occurs, a delay to the Completion Date is assessed as the length of time that planned Completion is later than planned Completion as shown on the Accepted Programme (Clause 63.5).

    It is vital that as planned Completion can change on a regular basis, the Contractor shows a realistic planned Completion in each revised programme submitted for acceptance, but also that the Contractor and Project Manager comply with the time scales in the contract for assessing and implementing compensation events as it is not unusual for planned Completion to be shown later than the Completion Date, leading one to believe the Contractor is in delay, yet the reason for this is either that the time scale in the contract is being adhered to but is giving a false picture as a programme is submitted whilst the Contractor is still preparing its quotation for a compensation event, or that compensation events have not been assessed and implemented in accordance with the time scales stated in the contract.

- *the order and timing of the operations which the Contractor plans to do in order to Provide the Works*

    This shows the sequence, duration and dates of the various operations the Contractor plans to do, and should include dependencies and links between the various operations. It should also show sequence, duration and dates in

the procurement process, finalisation of information, placing orders, etc. It should also show work to be carried out off site.
- *the order and timing of the work of the Client and Others as last agreed with them by the Contractor or, if not so agreed, as stated in the Scope*

    The Contractor will state what it requires the Client and Others to do in order that the Accepted Programme can be met. These dates may already have been agreed with the Contractor, but may also have been inserted into the Scope by the Client. "Others" are Parties outside the contract, i.e. not the Client, Project Manager, Supervisor, Adjudicator, the Contractor or a Subcontractor or Supplier to the Contractor. Examples of "Others" would be planning authorities, utilities companies, etc.

    This requirement is often provided in the form of an Information Required spreadsheet identifying what information is required and the latest date for providing it, failing which a compensation event can occur (Clause 60.1(5)). It is important to recognise that the programme is not just the bar chart but may include spreadsheets, schedules and graphs.

- *the dates when the Contractor plans to meet each Condition stated for the Key Dates and to complete other work needed to allow the Client and Others to do their work*

    If there are Key Dates in the contract (Clause 11.2(11)), then the Contractor must show how and when it intends to meet each Condition by these Key Dates. There may be provision for other Contractors directly employed by the Client to carry out work, for example a fit out Contractor.

Provisions for

– *float*

    Float is any spare time within the Contractors programme, after time risk allowances have been included, and represents the amount of time that operations may be delayed without delaying following operations (free float) and/or planned Completion (total float). It can also represent the time between when the Contractor plans to complete and when the contract requires him to complete (terminal float).

    Float absorbs to a certain extent the Contractor's own delays or the delays caused by a compensation event, thereby lessening or avoiding any delay to planned Completion. In effect, no delay arises unless float on the relevant and critical operations reduces to below zero.

    Programming is never an exact science, so float gives some flexibility to the Contractor in respect of incorrect forecasts or its own inefficiencies. As the work progresses the float will change as output rates change, Contractor's risk events take place and also compensation events arise.

    The general belief, certainly amongst many Contractors, is that float belongs to them, as they wrote the programme, and therefore have the right to work to that programme and use any float that it contains. Conversely,

Clients believe that if it is free time then they have the right to use it, but it actually depends where the float is and what it is for.

In principle, float other than terminal float or time risk allowances is a shared resource, it is spare time, it belongs to the project, and therefore may be used by whichever Party needs it first.

There are three types of float:

(i) the amount of time that an operation can be delayed before it delays the earliest start of following operation (free float)
(ii) the amount of time that an operation can be delayed before it delays the earliest completion of the works (total float)
(iii) the amount of time between planned Completion and the Completion Date (terminal float).

Generally, with other contracts, if the Contractor shows that it plans to complete a project early, then it is prevented from completing as early as it planned, but it still completes before the contract completion date, then although there may be an entitlement to an extension of time under the contract, for example exceptionally adverse weather, none will be awarded as there is no delay to contract completion, but it may have a right to financial recovery subject to him proving loss and/or expense.

The ECC deals with this in a different way. If the Contractor shows on its programme planned Completion earlier than the Completion Date and it is prevented from completing by the planned Completion Date by a compensation event, then, when assessing the compensation event, Clause 63.5 states "a delay to the Completion Date is assessed as the length of time that, due to the compensation event, planned Completion is later than planned Completion as shown on the Accepted Programme current at the dividing date"; therefore, any terminal float is retained by the Contractor, the period of delay being added to the Completion Date to determine the change to the Completion Date.

Any delay to planned Completion due to a compensation event therefore results in the same delay to the Completion Date. Therefore, the extension is granted on the basis of time the Contractor is delayed, i.e. entitlement, not on how long it needs to achieve the current Completion Date.

Similarly, with Key Dates, "a delay to a Key Date is assessed as the length of time that, due to the compensation event, the planned date when the Condition stated for a Key Date will be met is later than the date shown on the Accepted Programme current at the dividing date".

In assessing any delays to Completion Date and/or Key Dates only those operations that the Contractor has not completed and that are affected by the compensation event are changed.

– *time risk allowances*

Time risk allowances are essentially a form of float and are included within the duration of a specific operation. However, time risk allowances are normally "within an operation" rather than "between operations".

Essentially, a time risk allowance is the difference between what is the most "optimistic" (shortest) duration to complete an operation and what is a "realistic" (expected) duration, bearing in mind the risks that the Contractor may face in completing that operation. It is important that the Contractor's staff are fully informed as to how the time risk allowances are being presented, so there is no misunderstanding of how the works have been programmed and priced.

The Contractor's time risk allowances are to be shown on its programme but there are no requirements in the contract as to how these allowances should actually be shown, three possible methods being:

(i) show a single bar for an operation, within which the time risk allowance is shown. This is difficult to do with some planning software
(ii) show a separate bar which represents the time risk allowance for a specific operation. This, however, adds to the lines on the programme
(iii) show a separate column identifying the time risk allowance included within the total duration of the operation.

Clause 63.8 states that the assessment of the effect of a compensation event includes risk allowances for cost and time for matters that have a significant chance of occurring and are at the Contractor's risk.

It follows that they should be retained in the assessment of any delay to planned Completion due to the effect of a compensation event. These allowances are owned by the Contractor as part of its realistic planning to cover its risks in pricing its tender and also in managing changes. They should be either clearly identified as such in the programme or included in the time periods allocated to specific activities.

It is important to recognise that time risk allowances belong to the Contractor and are retained in the assessment of any delay to planned Completion due to a compensation event.

– *health and safety requirements*
   This will include time taken in complying with contractual and statutory requirements, but also in inducting new staff and operatives.
– *the procedures set out in the contract*
   This can include time scales for acceptance of design proposals, proposed subcontractors, etc.
• *the dates when, in order to Provide the Works in accordance with its programme, the Contractor will need*
   – access to a part of the Site if later than its access date
   – acceptances
   – Plant and Materials and other things to be provided by the Client
   – information from Others.

Generally, the Contractor should be given access to parts of the Site, acceptance of its design, subcontractors and programme in a timely manner to allow him to Provide the Works. However, in respect of access, the Contractor may be in delay and may therefore not require access on the date shown in the contract; there may also be certain operations for which there may be a long lead in time, which the Client or Others may not appreciate and therefore the Contractor can include the requirement for acceptances and information at an early date.

This information may be highlighted as milestone on the bar chart itself, or attached to it in the form of information required schedules, resource schedules, etc.

Anything submitted by the Contractor as part of its programme must clearly and simply show the relevant operations for which access, Plant and Materials or information are required, and the implications if they are not provided.

It is important that the Contractor and the Project Manager then manage the process by raising any concerns in respect of their provision as early warnings.

- *for each operation, a statement of how the Contractor plans to do the work identifying the principal Equipment and other resources which will be used.*

The Contractor is required to provide a statement of how it intends to carry out each and every item on each programme it submits for acceptance. Note that this is not a method statement in the full sense, which would include matters of health and safety, environmental precautions, etc.

Whilst accepting that the contract requires this information, this can be a somewhat idealistic clause that requires the Contractor to comply with the daunting and in many cases somewhat unreasonable requirement to produce all the information in its first programme submitted for acceptance. However, particularly on a large complex project with a long duration, the Contractor may not yet know how it will carry out each operation and what resources it will use. It also may not have appointed a Subcontractor for a part of the works so will not have method statements from the company who is actually carrying out the works.

It is also a significant task for the Project Manager to accept all the information the Contractor has submitted, so it may be appropriate in this respect, for the Contractor and the Project Manager to agree a rolling programme of submission of method statements as the work progresses and the methods and resources may become clearer, perhaps each submission covering the next three or four months' operations.

Whilst Clause 31.2 does not specifically call for method statements, such statements have become increasingly important in the construction industry, particularly as health and safety legislation makes the Parties more responsible for stating the methodology they are going to use in carrying out the works, and how they have provided for the various risks involved. For a programme to be meaningful, it should always be linked to method statements so that the methodology of carrying out the works and also the resources the Contractor intends to use are clearly stated.

The contract does not specify the form these statements should take and in many cases a Contractor will spend a great deal of time and paper producing detailed statements, which can be almost meaningless. They may be presented as a written statement or in tabular form but they should clearly show how the Contractor intends to do the work, main headings being:

1. Title of each operation
2. Description of method
3. Quantities (where relevant)
4. Principal Equipment to be used with outputs and durations
5. Labour type and gang sizes with outputs and durations
6. Plant to be installed
7. Materials to be used.

NB: Cost is not normally included in method statements.

When preparing method statements, the Contractor should always consider and show:

1. What is to be done
2. How it is going to be done
3. Who is going to do it
4. Where it is to be done
5. When it is to be done.

This information will then initially allow the Project Manager to consider whether the Contractor is providing sufficient resources to Provide the Works within the time available and also safely, but also, as the programme is revised and methods of construction and resources used may change, the Project Manager is then kept up to date as to the Contractor's intentions.

This information will also prove invaluable when assessing compensation events.

- *other information which the Scope requires the Contractor to show on a programme submitted for acceptance.*

Again, this shows that the Scope is not just the drawings and specifications, but information on all that the Contractor is required to do in Providing the Works.

From these requirements, one can see that the programme under an ECC contract is not just the bar chart, but a collection of documents including method statements, risk assessments, resource analyses, etc. All of these are submitted by the Contractor as its programme, and all of these must be considered by the Project Manager in determining whether to accept the programme. Extreme care should be taken to only require documentation that has a purpose and use in the administration.

It is essential that the Contractor either submits a programme with his tender, or within the time scale specified within Contract Data Part 1, and showing the information that the contract requires.

Failure to do so will entitle the Project Manager, under Clause 50.5, to retain one quarter of the Price for Work Done to Date in his assessment of the amount due.

Note that the amount is only withheld if the Contractor has not submitted a first programme which shows the information that the contract requires, e.g. method statement, time risk allowances, etc.

If the Contractor has submitted a programme that contains all the information that the contract requires, but the Project Manager disagrees with, for example, part of the method statement or the programme has not yet been accepted, then the provision, and the associated amount retained, does not apply.

Note also that, in respect of compensation events, under Clause 64.1:

- *if, when the Contractor submits quotations for the compensation event, it has not submitted a programme or alterations to a programme which the contract requires it to submit; or*
- *if, when the Contractor submits quotations for the compensation event, the Project Manager has not accepted the Contractor's latest programme for one of the reasons stated in the Contract.*

the Project Manager may assess a compensation event himself.

Clause 31.3 requires that within two weeks of the Contractor submitting a programme for acceptance, the Project Manager either accepts or does not accept the Contractor's programme, and if it does not accept it, the Project Manager clearly states why it does not accept it.

There is no "implied acceptance" if the Project Manager fails to accept or not accept the Contractor's first programme or subsequent submitted programmes. If it does not accept the programme, and it is not for a reason stated in the contract that is in Clause 31.3, then it is a compensation event, though the Contractor is not prevented from commencing the works so it is difficult to envisage what the Contractor would be seeking to recover from the breach through the compensation event process?

It has been known for Project Managers to respond to the submission of the programme either by making no comment at all within the two weeks they have to respond, or by stating that the programme is accepted on condition that the Contractor makes certain amendments to it. Neither of these are either acceptances or non acceptances.

Because of these issues of non-response, or at least vague responses, the NEC4 ECC now has a new provision within Clause 31.3 in that, if the Project Manager fails to reply to the Contractor's submission within the time allowed, the Contractor may notify to that effect. If the failure to respond continues for a further one week after the Contractor's notification, it is treated as acceptance by the Project Manager.

Note: The notice from the Contractor is the precondition to deemed acceptance by the Project Manager.

This will hopefully goad an "errant" Project Manager to do what the contract requires it to do!

Question 3.9 How often must the Contractor in an NEC4 Engineering and Construction Contract submit a revised programme?

The timing for submission of revised programmes for acceptance is defined by Clause 32.2:

- *within the period for reply after the Project Manager has instructed the Contractor to*
    For example, the Project Manager could instruct the Contractor to submit a programme to enable an issue to be discussed in an early warning meeting.
- *when the Contractor chooses to and, in any case*
    Again, the Contractor may feel it beneficial to submit a programme to be discussed in an early warning meeting.
- *at no longer interval than the interval stated in the Contract Data from the starting date until Completion of the whole of the works*
    This is the longest period for submission of a revised programme.

The revised programme shows:

- the actual progress achieved on each operation and its effect upon the timing of the remaining work
    The progress achieved may be shown on the programme itself, or appended to it in the form of a progress report. Operations which are not yet completed will normally be shown as percentage complete, being the progress achieved to date against the total work within the operation on the date the assessment is made. It is useful if the Contractor and the Project Manager agree the progress statement before the revised programme comes into being. The submitted revised programme is then an undisputed statement of fact.

    The revised programme should reflect "planned v actual" progress including early or delayed start, early or delayed completion of each operation. It is important that, if the Contractor states a percentage completed to date, he should show actual progress achieved rather than time expired to date. Also, the Contractor should not just make statements such as "operation on programme" as this may be misleading.

    An operation on programme does not necessarily mean that the operation is not delayed; for example, the Contractor is expecting to complete an operation late, then stating the words "operation on programme" can be interpreted as either in accordance with the contract, or progressing late in accordance with his expectations.

It is also important to recognise that many operations may not progress on a "straight line" basis, for example building a brick wall may initially involve work and resources in setting out and building corners, with the bulk areas following on.

- how the Contractor plans to deal with any delays and to correct notified Defects

  The Contractor should show in his revised programme any delays that may or may not be caused by him, and also time needed to correct his own Defects.

- any other changes that the Contractor proposes to make to the Accepted Programme

  An example of this could be that the Contractor resequences part of the work or proposes to use a different method or different Equipment to that specified by him in a previous method statement.

Note that NEC3 required the Contractor to show "the effects of implemented compensation events".

The omission of this bullet point in the NEC4 Engineering and Construction Contract is welcomed, as a compensation event is not "implemented" until the quotation has been accepted or assessed by the Project Manager.

Whilst the completion date is not changed until a compensation event has been implemented, there is a danger in only showing compensation events that have been implemented as the revised programme is not reflecting reality.

Example

The Project Manager instructs the Contractor not to paint the walls to certain rooms of a new building as he is considering an alternative finish which he will instruct at a later date. The following week, the Contractor is due to submit a revised programme, but he has only just submitted his quotation to the Project Manager for the omission of the painting, so there is probably a couple of weeks before the Project Manager replies and the compensation event is implemented. Should the Contractor in the meantime be showing the walls as to be painted or not to be painted as the compensation event has not yet been implemented? It should show the walls as not to be painted as it reflects reality.

It is therefore very much a "living" programme. If the Contractor does not submit a programme that the contract requires, then the Project Manager can assess compensation events, without receiving a quotation from the Contractor (Clauses 64.1 and 64.2).

The Accepted Programme:

- effectively provides an agreed record of the progress of the job and where the delays have come from
- provides a realistic base for future planning by both the Contractor and Project Manager
- is the base from which changes to the Completion Date are calculated
- is the base from which additional costs are derived. Because each operation has a method statement and resources attached to it, the change in resources can be calculated and hence the change in costs. Because information is so much more transparent, there is more scope for working together.

Question 3.10 In an NEC4 Engineering and Construction Contract being carried out under Option A (priced contract with activity schedule), can we require the Contractor to bring forward Completion if the Client has a need to take over the project earlier?

Acceleration in many contracts normally means increasing resources, working faster, or working longer hours so that completion can be achieved by the Completion Date; in effect, the Contractor is catching up to recover a delay. However, within the ECC acceleration means bringing increasing resources, working faster, or working longer hours so that Completion can be achieved *before* the Completion Date.

The ECC has always had provision for acceleration, but within the NEC4 ECC (Clause 36), there has been a redraft of the provision to give a more collaborative approach.

In the NEC3 ECC, the Project Manager may instruct the Contractor to submit a quotation for acceleration. It also stated any changes to any Key Dates to be included within the quotation.

In the NEC4 ECC, the Contractor and the Project Manager may propose an acceleration to achieve Completion before the Completion Date. If they mutually agree to consider the proposed change, the Project Manager instructs the Contractor to submit a quotation for acceleration. It also states any changes to any Key Dates to be included within the quotation.

In reality, the decision to consider acceleration is usually a collaborative decision between the Project Manager and the Contractor, before the Contractor is instructed to submit a quotation for acceleration, rather than the Project Manager unilaterally issuing an instruction to the Contractor to submit a quotation, so the new provision mirrors reality.

The Contractor provides a quotation within three weeks of being instructed to do so.

The quotation must include proposed changes to the Prices, and a programme showing the earlier Completion Date and Key Date. As with quotations for compensation events, the Contractor must submit details within its quotation; however, the contract does not prescribe how the quotation is to be priced, e.g. based on the Schedule of Cost Components.

The Project Manager replies to the quotation within three weeks of receiving it. The reply may be either:

- a notification that the quotation is accepted or
- a notification that the quotation is not accepted and that the Completion Dates and Key Dates are not changed.

Note that there is no remedy if the Contractor does not submit a quotation, or if the Contractor's quotation is not accepted by the Project Manager, in that it could make its own assessment, as it could with a compensation event. Acceleration can only be undertaken by agreement between the Project Manager and the Contractor and cannot be imposed on the Contractor or any assessment imposed upon him without its agreement.

When the Project Manager has accepted the quotation for acceleration, it changes the Prices, the Completion Date and any Key Dates and accepts the revised programme.

Chapter 4

Quality management

The roles of the Parties, testing requirements, searching for Defects, etc.

Question 4.1 We have identified a number of Defects on a project which a Contractor is carrying out under the NEC4 Engineering and Construction Contract, but the Contractor denies that they are Defects. How do the NEC4 contracts define a Defect?

A Defect is defined within the contract (Clause 11.2(6)) as:

- *a part of the works which is not in accordance with the Scope* or
- *a part of the works designed by the Contractor which is not in accordance with the applicable law or the Contractor's design which the Project Manager has accepted.*

In the first bullet, the Scope provides the reference point for what the Contractor has to do to Provide the Works. If the work done by the Contractor does not comply with the Scope, then unless the Defect is accepted (Clause 45), the Contractor is obliged to correct the Defect.

In most cases the work done will fall short of the Scope, but as the words "not in accordance with" are used, it is a compliance issue, therefore work that *exceeds* the requirement within the Scope would also be a Defect. Clearly, the Project Manager may wish, having discussed with the Client, to accept such a Defect!

A Defect may not necessarily mean that the work is not fit for purpose in that the Defect could simply be a colour variation; for example, the Scope stated a certain shade of red paint and the Contractor used a different shade of red paint, in which case a Defect exists as that part of the works was not in accordance with the Scope. In such a case, unless the shade of red is a particular corporate colour or a stipulation of a regulatory authority it may be prudent to accept the Defect.

It is also possible that work is "defective" but as it complies with the Scope, it is not a Defect! For example, the specification within the Scope may stipulate use of a particular material which then fails. In that respect, the work is defective but as it complies with the Scope, it is actually not a Defect.

In the second bullet, if the Project Manager accepts the Contractor's design under Clause 21.2, then subsequently the Contractor either does not comply with the applicable law, or changes the design, then again that is a Defect. If the Contractor wishes to change his design then he has to re-submit the new design to the Project Manager for his acceptance.

Until the defects date, the Supervisor is obliged to notify the Contractor of every Defect as soon as he finds it, and the Contractor is obliged to notify the Supervisor as soon as he finds it (see Figure 4.1).

Whilst the requirement for the Supervisor to notify the Contractor of each Defect is useful, as the Contractor may not have been aware of the Defect or that it was a Defect, so he can correct it immediately, the obligation for the Contractor to notify the Supervisor of each Defect as soon as he finds it is somewhat cumbersome as the Supervisor probably does not want or need to know about every Defect, particularly if the Contractor is already correcting it.

Question 4.2 **On an Engineering and Construction Contract using Option C (target contract with activity schedule), the Contractor has advised us that he is entitled to be paid for correcting a Defect. We disagree that we should be paying him for correcting Defects.**

The Contractor is liable for correction of Defects and as a general rule, he is not entitled to be paid additional monies or given additional time in which to correct those Defects.

However, note that under Options C, D and E where the Contractor is reimbursed his Defined Cost, two Disallowed Costs under Clause 11.2(26) deal specifically with Defects.

The first bullet point states:
the cost of:

- *correcting Defects after Completion*

If the Contractor corrects Defects *after* Completion then he *is not* reimbursed the Defined Cost of doing so; however, and this is where Clients and many NEC4 practitioners express some concern, if the Contractor corrects Defects *before* Completion then it *is* reimbursed the Defined Cost of doing so.

The threshold is when the Defect is corrected, not when it occurs, so a Defect that appears before Completion, but is corrected after Completion is a Disallowed Cost.

There is no limit to the type or size of Defects, or why the Defect occurred, e.g. through the carelessness of the Contractor.

Whilst some may see this as the Contractor's right to repeatedly attempt to get the work right at the Client's expense, it must be remembered that in the case of Option C and D these are target contracts, therefore if the Contractor is paid for correcting a Defect, not only is this bad for his reputation, but as it is not a compensation event he is not being given additional time to correct the Defect and the additional cost paid will reduce his entitlement to the Contractor's share at Completion.

Contract:	DEFECTS NOTIFICATION
Contract No:	D/N No..
The Contractor is notified of the following defects:	
Description	**Date Corrected**

The defect correction period is:

Certified all defects corrected..............................(Supervisor)   Date: ....

Copied to:

Contractor ☐   Project Manager ☐   Supervisor ☐   File ☐   Other .... ☐

Figure 4.1 Defects notification

The second bullet point states:
the cost of:

- *correcting Defects caused by the Contractor not complying with a constraint on how he is to Provide the Works stated in the Scope*

So, if the Contractor has to correct a Defect because he did not comply with the Scope, whether it is before or after Completion, then he will not be reimbursed the cost of doing so.

Question 4.3   What is the meaning of the term "search for a Defect" under Clause 43.1 of the NEC4 Engineering and Construction Contract?

Under Clause 43.1, the Supervisor (not the Project Manager, unless the Supervisor has delegated the authority to him) has the authority to instruct the Contractor to search. This action is normally defined as "uncovering", "opening up" or "uncovering" in other contracts. It is usually required in order to investigate whether a Defect exists, and possibly the cause, and the corrective measures required. Searching can include uncovering, dismantling, reassembly, providing materials and samples and additional tests which the Scope did not originally require.

If the Supervisor instructs the Contractor to search for a Defect and no Defect is found, this is a compensation event (Clause 60.1(10)). However, if the search is needed only because the Contractor gave insufficient notice of doing work obstructing a required test or inspection, then it is not a compensation event.

Question 4.4   What can be done on an NEC4 Engineering and Construction Contract if the Contractor fails to correct Defects?

The Project Manager arranges for the Client to allow access to parts of the works taken over in order to correct a Defect.

If the Contractor does not correct the Defect within the defect correction period in the contract, the Client assesses the cost of having the Defect corrected by other people and the Contractor pays this amount.

Note that the Client may choose not to have the Defect corrected by anyone, but that is his prerogative, the Contractor still pays this amount.

Question 4.5   When is the Defects Certificate issued under an NEC4 Engineering and Construction Contract, and who issues it?

Quality management   129

The Defects Certificate is issued by the Supervisor to the Contractor, and to the Project Manager and the Client, at the defects date if there are no notified Defects, or otherwise at the earlier of

- *the end of the last defect correction period and*
- *the date when all Defects have been corrected.*

(See Figure 4.2.)

Whilst many contracts require any Defects to be corrected before the Defects Certificate (or its equivalent) is issued, under the Engineering and Construction Contract the Defects Certificate is issued by the Supervisor at the appropriate time as stated within the contract, and may either list Defects that the Contractor has not corrected, or include a statement that there are no Defects.

Some practitioners state that as the Project Manager issues the Completion Certificate, he should also issue the Defects Certificate, which is common with most other contracts that have an Engineer or an Architect/Contract Administrator, but by requiring the Supervisor to issue the Defects Certificate, the Engineering and Construction Contract reinforces the role and responsibility of the Supervisor in dealing with Defects.

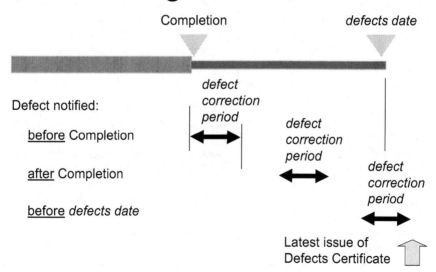

Figure 4.2 Defects correction period and the Defects Certificate

Note that if Option X16 (Retention) has been selected, under Clause X16.2, the amount retained is halved

- *in the assessment made after Completion of the whole of the works or*
- *in the next assessment after the Client has taken over the whole of the works if this is before Completion of the whole of the works.*

The amount retained remains at this amount until the Defects Certificate is issued. No amount is retained in the assessments made after the Defects Certificate has been issued.

Note that this may mean that the retention is released to the Contractor, even though there may still be outstanding Defects yet to be corrected.

If stated in the Contract Data, or if agreed by the Client, the Contractor may give the Client a retention bond for the total amount to be retained. If the retention bond is accepted, any amount retained is then paid to the Contractor in the next assessment.

The Defects Certificate is not conclusive, in that if a Defects Certificate states that there are no Defects it does not prevent the Client from exercising its rights should a defect arise later, probably as a latent defect, or the Supervisor did not find or notify it.

However, the Client may wish to take action against a Supervisor who did not carry out the level of inspection he had paid him to do!

The Supervisor will normally only list what are usually defined as "patent defects", i.e. those that are observable from reasonable inspection at the time, examples being a defective concrete finish or an incorrect paint colour, and will not include what are usually defined as "latent defects", which may be hidden from reasonable inspection and may come to light at a later date, examples being some structural defects.

Note that the Contractor's liability for correction of latent defects and other costs associated with them will be dependent on the applicable law, and liability is likely to remain despite the issue of the Defects Certificate.

Question 4.6   Our project, which we are about to carry out using the NEC4 Engineering and Construction Contract Option B (priced contract with bill of quantities), will be on completion a very high security building with restricted access. Once the Client has taken over the works, we cannot allow the Contractor to return to correct any Defects. How do we then deal with any Defects that may arise?

Normally in a construction contract, the Contractor is obliged to correct Defects arising in the works, during the works and following Completion, in the case of NEC4 contracts until the defects date. This is an obligation, but also it gives the Contractor the opportunity to correct any Defects.

However, there may be a situation where the Client cannot allow the Contactor back into the works due to high security e.g. nuclear projects.

Under Clause 46.2 of the Engineering and Construction Contract, if the Contractor is not given access in order to correct a notified Defect before the *defects date*, the Project Manager assesses the cost to the Contractor of correcting the Defect and the Contractor pays this amount. The Scope is then treated as having been changed to accept the Defect.

This is an unusual clause in that most contracts as stated provide for the Contractor to correct his own Defects, if he does not take the opportunity he is liable for the cost of someone else correcting it.

In this case, as the Contractor is denied the right to re-enter the building he is subject to the Project Manager's subjective opinion as to how much cost the Contractor would have incurred in correcting the Defect(s) if he had been allowed access.

Whilst this may seem a simple and effective solution, in practical terms it is very difficult for the Project Manager to assess the cost to the Contractor of correcting the Defect.

Question 4.7 The Contractor has installed the wrong suspended ceiling tiles to the training rooms within our new building; however, we urgently need to take over the works and cannot wait for him to correct the Defect. How can we manage this issue?

A Defect is defined as "a part of the works that is not in accordance with the Scope" or "a part of the works designed by the Contractor which is not in accordance with the applicable law or the Contractor's design which the Project Manager has accepted" (Clause 11.2(6)).

In the event that a Defect becomes apparent, there are two choices:

(i) The Contractor corrects the Defect so that the part of the Works is in accordance with the Scope or the design is in accordance with the applicable law or the Contractor's design.
(ii) The Contractor and Project Manager may each propose to the other that the Scope should be changed so that a Defect does not have to be corrected (Clause 45). Note that "each proposes to the other" requires the acceptance of the Contractor and the Project Manager, the latter probably discussing the matter with the Client.

Example

The Contractor has carried out a large area of wall tiling in a proposed new railway station. Whilst visually the quality of the work appears to be very good, the tiles have not been laid to the required tolerance within the Scope and therefore the work is defective.

(continued)

(continued)

The railway station is due to be completed in two weeks, so if the Contractor has to take down the wall tiles, re-order another batch of tiles and then lay the new tiles to the required tolerance, that could take a considerable time and possibly prevent completion of the works or that specific part of the works.

In this case, provided both the Contractor and the Project Manager are prepared to consider changing the Scope, then the Contractor provides a quotation to the Project Manager stating a financial saving (reduced Prices), based on him not having to correct the Defect or an earlier Completion Date or both, and if the Project Manager accepts the quotation then the Scope is changed and the Prices and/or Completion Date are changed in accordance with the quotation.

This is obviously not an option where the design is not in accordance with the applicable law. If the Contractor and Project Manager consider this change, the Contractor submits a quotation for reduced Prices or an earlier Completion Date, or both.

If the Project Manager accepts the quotation he gives an instruction to change the Scope, the Prices and the Completion Date.

Chapter 5

Payment provisions

Payment and non-payment under the various options, use of Disallowed Cost, etc.

Question 5.1   We wish to make use of a Project Bank Account. Can we do this with an NEC4 contract?

In 2008, the Office of Government Commerce (OGC) published a guide to fair payment practices, following which the NEC Panel prepared a document in June 2008 to allow users to implement these fair payment practices into NEC contracts.

This is now Option Y(UK)1, which provides for a Project Bank Account to be set up that receives payments from the Client which is in turn is used to make payments to the Contractor and Named Suppliers. The Project Bank Account is established within three weeks of the Contract Date.

There is also a Trust Deed between the Client, the Contractor and Named Suppliers containing the necessary provisions for administering the Project Bank Account.

The Contractor should include in any subcontracts for Named Suppliers to become Party to the Project Bank Account through a Trust Deed via a Joining Deed.

The Contractor notifies the Named Suppliers of the details of the Project Bank Account and the arrangements for payment of amounts due under their contracts.

The Named Suppliers will be named within the Contractor's tender, but also additional Named Suppliers may be included subject to the Client's acceptance by means of a Joining Deed, which is executed by the Client, the Contractor and the new Named Supplier. The new Named Supplier then becomes a Party to the Trust Deed.

As the Project Bank Account is maintained by the Contractor, he pays any bank charges and also is entitled to any interest earned on the account. The Contractor is also required at tender stage to put forward his proposals for a suitable bank or other entity which can offer the arrangements required under the contract.

The process every month is that the Contractor submits an application for payment including details of amounts due to Named Suppliers in accordance with their contracts.

The Client makes payment to the Project Bank Account, the Contractor makes payment to the Project Bank Account of any amounts which the Client has notified the Contractor intends to withhold from the certified amount and which is required to make payment to Named Suppliers.

The Contractor then prepares the Authorisation, setting out the sums due to Named Suppliers. After signing the Authorisation, the Contractor submits it to the Client for signature and submission to the project bank.

The Contractor and Named Suppliers then receive payment from the Project Bank Account of the sums set out in the Authorisation after the Project Bank Account receives payment.

In the event of termination, no further payments are made into the Project Bank Account.

Question 5.2 We note that the NEC4 Engineering and Construction Contract does not include provision for retention on payments to the Contractor. Why is this, and how can we include for retention?

It is true that retention is not included within the core clauses of the NEC4 Engineering and Construction Contract, but it is included as a Secondary Option, i.e. Option X16.

This is because not all clients will want to deduct retention from payments; therefore it is an option, rather than a core clause requirement.

The Client may retain a proportion of the Price for Work Done to Date once it has reached any retention free amount, the retention percentage and any retention free amount being stated in Contract Data Part 1.

Within Schedule of Cost Components Clause 4, which relates to Options C, D and E, and within Option F (Clause 11.2(25)), Defined Cost does not include amounts deducted for retention, so, if the Contractor deducts retention from a Subcontractor, the figure before the deduction is used in calculating Defined Cost, ensuring that there is no "double deduction" of retention.

Following Completion of the whole of the works, or the date the Client takes over the whole of the works, whichever happens earlier, the retention percentage is halved, and then the final release is upon the issue of the Defects Certificate.

It is important to note that retention is held against undiscovered defects and not incomplete work.

If stated in the Contract Data, or if agreed by the Client, the Contractor provides a retention bond provided by a bank or insurer, accepted by the Project Manager, and in the form stated in the Scope.

Question 5.3 We need to pay the Contractor on an NEC4 Engineering and Construction Contract, partly in US dollars and partly in the local currency of the country in which the project is located. How can we do this?

It is a fairly common requirement in international contracts for part of the payments to the Contractor or the Consultant to be in, for example, the US dollar, which may be linked to oil prices, and part in the local currency of the country in which the project is located.

The currency of the contract is stated in Contract Data Part 1, for example, the project may be in Qatar, and the currency of the contract would most probably be the Qatari Riyal, in which case all payments would be made in Qatari Riyals.

Option X3 provides for items or activities to be paid in an alternative currency, the items or activities, the currency and the total maximum payment in this currency to be listed in the Contract Data, beyond which payments are made in the currency of the contract.

The exchange rates, their source and date of publication are also referred to in the Contract Data.

Question 5.4 We wish to pay the Contractor on an NEC4 Engineering and Construction Contract a mobilisation payment equating to 10% of the value of the contract. How can we do this?

Mobilisation payments are appropriate when the Contractor will incur significant "up front" costs before he starts receiving payments, for example in pre-ordering specialist Materials, Plant or Equipment.

The NEC4 Engineering and Construction Contract provides for such payments through Secondary Option X14 (Advanced Payment to the Contractor).

The payment is made within the first payment assessment or if an Advanced Payment Bond is required, at the next assessment after the Client receives the bond.

If an advanced payment bond is required it is issued by a bank or insurer which the Project Manager has accepted, the bond being in the amount that the Contractor has not repaid.

Advance payment bonds provide security when an advance payment is made to a Contractor for works to be performed. The Project Manager must accept the provider of that bond.

The amount of the repayment instalments is stated in Contract Data Part 1.

Question 5.5 What is the Fee in an NEC4 Engineering and Construction Contract intended to cover?

Fee is defined under Clause 11.2(10) as *"the amount calculated by applying the fee percentage to the amount of Defined Cost"*.

Whilst many practitioners believe that the Fee simply covers "overheads and profit", the definition is a little wider. All components of cost not listed in the Schedules of Cost Components are covered by the fee percentage.

There is no schedule of items covered by the fee percentage, but the following list, whilst not exhaustive, gives some examples of cost components not included in the Schedules of Cost Components.

136 Payment provisions

1 Profit
2 The cost of offices outside the Working Areas, e.g. the Contractor's head office
3 Insurance premiums
4 Performance bond costs
5 Corporation tax
6 Advertising and recruitment costs
7 Sureties and guarantees required for the contract
8 Some indirect payments to staff.

From this one can see that there are some elements covered by the percentage for Fee which could be mistakenly assumed as being covered by the Schedules of Cost Components.

Prior to the NEC3 Engineering and Construction Contract, there was only one fee percentage in the contract, which was applicable to the Defined Cost.

Then the NEC3 Engineering and Construction Contract provided for the Contractor to tender two fee percentages:

1 the *subcontracted fee percentage applied to the Defined Cost of* subcontracted work
2 the *direct fee percentage* applied to the Defined Cost of other work.

It was essential that these two fee percentages were correctly allocated to the appropriate costs when assessing payments and compensation events, though in reality, many Contractors tended to bracket the two fee percentages together as a single fee percentage, which they were entitled to do, and this in effect made life easier for the Contractor and the Project Manager when assessing payments and compensation events.

The NEC4 Engineering and Construction Contract has now reverted back to one Fee percentage in the contract, which is applicable to the Defined Cost.

Question 5.6 The Contractor on an NEC4 Engineering and Construction Contract has failed to submit a programme to the Project Manager. What can we do to remedy this?

First, in respect of payments due to the Contractor, if no programme is identified in Contract Data Part 2 (i.e. it was submitted with the Contractor's tender), one quarter of the Price for Work Done to Date is retained (Clause 50.5) until the Contractor has submitted the first programme showing the information the contract requires.

Note that a Contractor who submits a first programme which shows the required information, but the Project Manager does not accept it, would not be liable for withholding of payment under this clause. The clause relates to the

Contractor's *failure* to submit a programme showing the information the contract requires, not the *acceptance* of that programme.

Second, in respect of compensation events, under Clause 64.1:

- *if, when the Contractor submits quotations for the compensation event, it has not submitted a programme or alterations to a programme which the contract requires it to submit; or*
- *if, when the Contractor submits quotations for the compensation event, the Project Manager has not accepted the Contractor's latest programme for one of the reasons stated in the Contract.*

The Project Manager may assess a compensation event himself.

Ironically, under Clause 64.1, although the Contractor is in the hands of the Project Manager in terms of assessment of a compensation event, it does give the Project Manager extra work in having to make that assessment due to the Contractor's failure.

Question 5.7 On an NEC4 Engineering and Construction Contract Option A (priced contract with activity schedule), if the Contractor has completed an activity, should he be paid for that activity if it contains Defects?

The Price for Work Done to Date (PWDD) under Option A is the total of the Prices for completed activities (Clause 11.2(29)): "*A completed activity is one without notified Defects the correction of which would either delay or be covered by immediately following work*". So, for example, if the activity was to install a 30-metre length of copper pipework on and including supports, attached to a wall, the pipework to remain exposed after installation, and the pipework and the supports were all installed but there was a Defect in one of the supports, then provided it does not delay or is covered by immediately following work, it is a completed activity, but with a Defect that needs to be corrected.

The Client paying for the activity because the Project Manager assesses that it is completed is not the Client accepting the work or the Defects within it, it is just a means to create cash flow to the Contractor. The Contractor is still obliged to correct the Defect and is also responsible if, for example, someone damages the work once it is paid for.

Conversely, if the pipework was a drain pipe in a trench and there is a Defect that will be covered up the following day when the Contractor backfills the trench, then it is not paid at the time of assessment; the Contractor must correct the Defect first.

The Contractor is not entitled to be paid for activities that are not complete or are only partly complete so the decision always has to be made by the Project Manager at the assessment date as to whether the activity is either complete or not complete

138 Payment provisions

Assessing the Contractor's right to payment under the contract is therefore different to the traditional approach of valuing the works carried out to date.

Question 5.8   We are Contractors tendering for a series of NEC4 Engineering and Construction Contracts using Option A (priced contract with activity schedule). How should we price our Preliminaries costs within the activity schedule so that we are paid correctly?

It is not uncommon for Contractors who are new to NEC contracts, when they submit tenders for an NEC4 Engineering and Construction Contract Option A (priced contract with activity schedule), to include their Preliminaries costs as a single sum of money within a single activity, believing that they will actually be paid these costs in a traditional way, i.e. in the first month a lump sum payment for setting up their Site establishment, in continuing months payments for maintaining Site as a time related sum, then in the final month a lump sum payment for clearing the Site; however, this is an incorrect assumption.

The Price for Work Done to Date (PWDD) under Option A is the total of the Prices for completed activities (Clause 11.2(29)); therefore the Contractor should consider breaking down his Preliminaries as follows:

1   Fixed preliminaries

These would comprise Preliminaries that may be single items, not directly influenced by progress on Site.

Typical examples are delivery of temporary accommodation and other equipment to Site, connection of telephones and other temporary services.

These should be identified within the Activity Schedule as single activities, e.g. "delivery of Site accommodation", or a group of activities identified as "set up Site".

When the activities are completed, they are included within the next assessment following completion.

2   Time related preliminaries

These comprise Preliminaries that are based on time on Site rather than having any direct relationship to the quantity of work carried out.

Typical examples are site management salaries, site accommodation hire charges, and scaffolding.

These should be identified within the Activity Schedule as time-based activities, e.g. "one month hire of temporary accommodation". Again, when the activities are completed, they are included within the next assessment.

On a similar theme, Contractors who are tendering for NEC4 Engineering and Construction Contracts Option A (priced contract with activity schedule), where they have design responsibility, often include their Design costs as a single sum.

The Contractor should not just include an activity titled "Design" as there are many aspects of design, and one could argue that design is not completed until the project is completed, so he could deny himself payment for a considerable period.

Ideally, the design related activities should be linked to various deliverables rather than generic activity descriptions such as "prepare drawings".

The various elements of design could be related to, for example, the various stages within the RIBA Plan of Work, or in its simplest form there could be separate activities for the design of various elements of the project, and for the granting of planning approval, etc.

Question 5.9   On an NEC4 Engineering and Construction Contract Option B (priced contract with bill of quantities), the Contractor contends that if the quantities change, he is entitled to revise his rates. Is that true?

No, that is incorrect. Under Option B, if the quantities change, and the change is not caused by a compensation event, the original rates apply.

However, within Option B, one must also consider the following clause:

Clause 60.4

> A difference between the final total of work done and the quantity for an item in the Bill of Quantities is a compensation event if:

- *the difference does not result from a change to the Scope*
- *the difference causes the Defined Cost per unit of quantity to change*
- *the rate in the Bill of Quantities for the item multiplied by the final total quantity of work done is more than 0.5% of the Total of the Prices at the Contract Date.*

A difference between the final total quantity of work done and the quantity for an item on the Bill of Quantities is not a compensation event in itself, unless it satisfies all three bullet points within the clause.

If we examine each of the bullet points in turn:

- *the difference does not result from a change to the Scope*

If it had resulted from a change to the Scope, then it would have been dealt with as a compensation event under Clause 60.1(1).

- *the difference causes the Defined Cost per unit of quantity to change*

The purchase price of materials may be affected by a change in the quantity, for example a price reduction or increase may be applicable when the number of units exceeds or falls below a certain amount. Or the cost of mobilisation or delivery could have changed as the same vehicle delivers fewer units.

- *the rate in the Bill of Quantities for the item multiplied by the final total quantity of work done is more than 0.5 per cent of the Total of the Prices at the Contract Date.*

This final bullet point compares the final extended price for that item in the Bills of Quantities with the Total of the Prices at the Contract Date (many contracts refer to this as the "Tender Sum" or the "Contract Sum". If the extended price is more than 0.5 per cent of the Total of the Prices at the Contract Date, and both of the previous bullet points are satisfied, then it is a compensation event.

This final bullet then excludes items of a "minor value" in the Bill of Quantities.

Example

- Original quantity of work Road Gullies 150 at £125 = £18,750
- Due to an error in measuring the road gullies when preparing the Bill of Quantities, the final quantity is only 120
- Final quantity of work Road Gullies 120 at £125 = £15,000
- The Total of the Prices at the Contract Date is £2,000,000
- 0.5% × £2,000,000 = £10,000, therefore, assuming that all three conditions are satisfied, then it is a compensation event.

Note: This Clause does not mean that the rates in the Bill of Quantities are changed, but that the Contractor can be compensated for the financial effect of the change in quantities.

Question 5.10 What are "accounts and records" as required under an NEC4 Engineering and Construction Contract Option C, D, E and F? The Contractor has stated that there are certain accounts and records of costs that he cannot make available to the Project Manager as they are confidential, and also in some cases would breach data protection law. How can we pay the Contractor these costs?

Under Clause 52.2 of Options C, D and E, the Contractor is required to keep the following accounts and records to calculate Defined Cost:

- accounts of payments of Defined Cost;
- proof that the payments have been made;

- communications about and assessments of compensation events for Subcontractors; and
- other records as stated in the Scope.

Under Clause 52.3 of Option F, the Contractor is required to keep the following accounts and records (Clause 52.3) to calculate Defined Cost

- accounts of payments to Subcontractors;
- proof that the payments have been made;
- communications about and assessments of compensation events for Subcontractors; and
- other records as stated in the Scope.

Clause 52.4 of Options C, D, E and F require the Contractor to allow the Project Manager to inspect the accounts and records at any time within working hours.

What would these records consist of? Essentially, all that the Contractor requires to prove every cost for which he is seeking reimbursement under the contract, i.e. invoices, orders, timesheets, receipts, etc.

The level of checking of the Contractor's accounts and records is at the Project Manager's discretion, some wish to examine each cost and its relevant back-up document, some wish only to select certain costs at random, others wishing to only carry out a cursory check of costs.

Some Project Managers say that if the Parties really are acting in a spirit of mutual trust and co-operation as the contract requires, then there should be no need for a detailed examination of the Contractor's accounts, though this probably needs a quantum leap of faith for many Project Managers and Clients!

It is important to establish the format on which accounts and records are to be presented by the Contractor to the Project Manager, particularly in respect of assessing payments, and this can be detailed in the Scope, the following alternative methods normally being employed:

1   The Contractor submits a copy of *all* of his accounts and records to the Project Manager on a regular basis.
2   The Contractor gives the Project Manager access to his accounts and records and the Project Manager takes copies of all the records he requires.

    Whilst not expressly stated within the contract, the Project Manager should not have to travel too far to inspect them. It has been known for accounts and records to be "available for inspection" in a different country to the Site, the Contractor stating that they were available for inspection at any time during normal working hours!

    Again, the location and availability of accounts and records could be detailed within the Scope.

3   The Contractor gives the Project Manager direct access to his computerised accounts and records through a password or PIN, the Project Manager can then take copies of the records he requires.

If we then consider the payment process within Option C:

The Price for Work Done to Date is the Defined Cost which the Project Manager forecasts will have been paid by the Contractor before the next assessment date plus the Fee, so the assessment is a combination of Defined Cost paid and yet to be paid by the Contractor.

For example, if the assessment date is 1 June, then costs that are forecast to be paid by 1 July are also included.

This is a change to pre NEC3 editions of the contract which provided for the Contractor only to be paid cost that he *had paid* at the assessment date, the change being introduced to assist the Contractor's cash flow.

Whilst the forecast may appear to be no more than a subjective guess, the Contractor has an up to date programme and also has probably received many of the invoices which he will be paying before the next assessment date.

In practice, however, many Clients are resistant to forecasting costs a month ahead and therefore tend to amend the contract through a Z clause to retain the previous pre NEC3 wording.

In assessing the amount due if the Contractor has paid costs in a currency different to the currency of the contract, the Contractor is paid Defined Cost in the same currency as the payments made by him. All payments are converted to the currency of the contract applying the exchange rates, in order to calculate the Fee.

In response to the question that the Contractor has stated that there are certain accounts and records of costs that he cannot submit to the Project Manager as they are confidential, and also in some cases would breach data protection law, the simple answer is that the Contractor cannot be paid costs which he cannot prove through accounts and records, as the Project Manager is unable to forecast that they will have been paid by the Contractor before the next assessment date as he has not been able to verify what these costs are.

The Contractor may feel that salaries in particular are confidential or sensitive, but he must make the accounts and records available to the Project Manager in order to be paid them; but in turn the Project Manager, as with all the accounts and records made available to him, must respect that confidentiality.

Question 5.11   What are Disallowed Costs and how are they deducted from payments due? Can Disallowed Cost be applied retrospectively to a payment made in a previous month? Is there a maximum time period for deducting Disallowed Cost?

Disallowed Costs are costs that would normally be payable to the Contractor, but are disallowed under certain clauses within the contract.

There are different clauses covering Disallowed Cost dependent on the selected main option:

Options C, D and E – Clauses C11.2.(26), D11.2.(26), E11.2.(26)

Disallowed Cost is cost which the Project Manager decides:

- *is not justified by the Contractor's accounts and records*

   The Contractor is obliged to keep accounts, proof of payments, communications regarding payments, and any other records as stated in the Scope (Clauses 52.3 and 52.3). If it does not have the accounts and records to prove a cost then the cost must be disallowed. It is not sufficient to merely prove that the goods or materials are on the Site for the Project Manager to see, or held off Site at a designated place for inspection, as with many other contracts.

- *should not have been paid to a Subcontractor or supplier in accordance with its contract*

   The Contractor is obliged to submit the proposed conditions of contract for each Subcontractor to the Project Manager for acceptance, unless an NEC contract is proposed, or the Project Manager has agreed that no submission is required.

   In addition, under Options C, D, E and F, the Contractor is required to submit the proposed contract data for each subcontract if an NEC contract is proposed and the Project Manager instructs the Contractor to make the submission. If the Contractor pays a Subcontractor or supplier an amount that is not in accordance with its contract with them, then this is disallowed.

- *was incurred only because the Contractor did not*

   - *follow an acceptance or procurement procedure stated in the Scope*

   The Scope may require the Contractor to comply with a specific procedure in respect of, for example, submission of design proposals or procurement of Subcontractors. If the Contractor does not comply with the procedure then associated costs are disallowed.

   - *follow an acceptance or procurement procedure stated in the Scope*

   If the Contractor incurs a cost that could have been avoided if the Contractor had given early warning, then it is disallowed.

   - *give notification to the Project Manager of the preparation for and conduct of an adjudication or proceedings of a tribunal between the Contractor and a Subcontractor or supplier*

and the cost of

- *correcting Defects after Completion*

   Correction of Defects after Completion is a Disallowed Cost. However, correction of Defects before Completion is not a Disallowed Cost. This is often a contentious issue with Clients and Project Managers objecting to paying the Contractor for correcting its own Defects!

However, one must consider this clause logically. The Contractor has quite possibly priced the risk of having to correct Defects which are its liability within its original tender, on an Option C contract for example, this price will be included within the target. Option C is a cost reimbursable contract until Completion, and the only way the Contractor can be paid for this risk, should it materialise, is through the Defined Cost process as Option C is a cost reimbursable contract until Completion. It is not a compensation event, so the target and/or the Completion Date are not changed. The Contractor is correcting the defect in its own time and potentially reducing the share it could make at Completion.

- *correcting Defects caused by the Contractor not complying with a constraint on how it is to Provide the Works stated in the Scope*

  A constraint may be stated in the Scope, such as a prescribed method of working. For example, the Scope may prescribe that a hardcore sub base filling must be rolled six times with a vibrating roller of a certain weight, but the Contractor does not do so, and later a Defect arises due to the Contractor's failure to comply with that stated constraint.

  If the Contractor does not comply with this constraint and a Defect occurs, then the Contractor's cost of correcting the Defect is disallowed.

- *Plant and Materials not used to Provide the Works (after allowing for reasonable wastage) unless resulting from a change to the Scope*

  Excess wastage of plant or materials beyond what is considered reasonable is a Disallowed Cost. The question of what is "reasonable" can often be debatable! As with all Disallowed Cost it is the Project Manager's responsibility to make the decision and to disallow the cost.

- *resources not used to Provide the Works (after allowing for reasonable availability and utilisation) or not taken away from the Working Areas when the Project Manager requested*

  This provision will include People and Equipment. If the Contractor does not remove Equipment when it is no longer required, then this is Disallowed Cost. If the Contractor is using more resources than it has planned and priced for, or its resources are inefficient, that is not a Disallowed Cost.

- *preparation for and conduct of an adjudication, payments to a member of the Dispute Avoidance Board or proceedings of the tribunal between the Parties.*

  If an adjudication, the involvement of the Dispute Avoidance Board, arbitration or legal proceedings occur, then each Party bears its own costs. This bullet was not introduced until NEC3 was published, so prior to NEC3 a Contractor could, in theory, refer a dispute to adjudication and whether or not it was successful, any costs arising could be included as cost and would not be disallowed.

Option F – Clause F11.2.(27)

Disallowed Cost is cost which the Project Manager decides:

- *is not justified by the Contractor's accounts and records*

  The Contractor is obliged to keep accounts, proof of payments, communications regarding payments, and any other records as stated in the Scope

(Clauses 52.3 and 52.3). If it does not have the accounts and records to prove a cost then the cost must be disallowed. It is not sufficient to merely prove that the goods or materials are on the Site for the Project Manager to see, or held off Site at a designated place for inspection, as with many other contracts.

- *should not have been paid to a Subcontractor or supplier in accordance with its contract*

  The Contractor is obliged to submit the proposed conditions of contract for each Subcontractor to the Project Manager for acceptance, unless an NEC contract is proposed, or the Project Manager has agreed that no submission is required.

  In addition, under Options C, D, E and F, the Contractor is required to submit the proposed contract data for each subcontract if an NEC contract is proposed and the Project Manager instructs the Contractor to make the submission. If the Contractor pays a Subcontractor or supplier an amount that is not in accordance with its contract with them, then this is disallowed.

- *was incurred only because the Contractor did not*

  – *follow an acceptance or procurement procedure stated in the Scope*

  The Scope may require the Contractor to comply with a specific procedure in respect of, for example, submission of design proposals or procurement of Subcontractors. If the Contractor does not comply with the procedure then associated costs are disallowed.

  – *give an early warning which the contract required him to give*

  If the Contractor incurs a cost that could have been avoided if the Contractor had given early warning, then it is disallowed.

  – *give notification to the Project Manager of the preparation for and conduct of an adjudication or proceedings of a tribunal between the Contractor and a Subcontractor or supplier or*

- *is a payment to a Subcontractor for*

  – *work which the Contractor states that the Contractor will do itself or*

  – *the Contractor's management*

- *and was incurred in the preparation for and conduct of an adjudication, or payments to a member of the Dispute Avoidance Board or proceedings of the tribunal between the Parties.*

  If an adjudication, the involvement of the Dispute Avoidance Board, arbitration or legal proceedings occur, then each Party bears its own costs. This bullet was not introduced until NEC3 was published, so prior to NEC3 a Contractor could, in theory, refer a dispute to adjudication and whether or not it was successful, any costs arising could be included as cost and would not be disallowed.

Clearly, with all the Disallowed Cost provisions, as the Project Manager is to decide that a cost is to be disallowed, then it has to be diligent enough to identify the cost, to quantify it, and to make the appropriate deductions form payments, which are not the easiest of tasks!

In answer to the question "can Disallowed Cost be applied retrospectively to a payment made in a previous month", the simple answer is yes, Disallowed Cost can be deducted from any payment that would otherwise be due to the Contractor and does not necessarily have to be within the current payment to which the Disallowed Cost relates.

Question 5.12 How is the Contractor's Share under the NEC4 Engineering and Construction Contract Option C (target contract with activity schedule) set and calculated?

The Client enters the Contractor's share percentage for each share range into Contract Data Part 1, the principle being that the Contractor receives a share of any saving and pays a share of any excess when Defined Cost plus Fee is compared with the target at Completion of the whole of the works and in the final payment following the issue of the Defects Certificate.

The Contractor tenders a price and includes an activity schedule in the same way as he would under an Option A contract. This price, when accepted, is then referred to as the "target".

The Contractor also tenders his percentage(s) for Fee. The original target is referred to as the "Total of the Prices at the Contract Date".

- The target price includes the Contractor's estimate of Defined Cost plus other costs, overheads and profit to be covered by his Fee.
- The Contractor tenders his Fee in terms of percentages to be applied to Defined Cost.
- During the course of the contract, the Contractor is paid Defined Cost plus the Fee.
- The target is adjusted for compensation events and also for inflation (if Option X1 is used).
- On Completion, the Project Manager assesses the Contractor's share in accordance with Clause 54.1 which, it has to be said, is at best a confusing clause, though the Guidance Notes clarify the clause! The Contractor then pays or is paid his share of the difference between the final total of the Prices and the final Payment for Work Done to Date according to a formula stated in the Contract Data.

This motivates the Contractor to decrease costs. Many refer to this sharing of risk and opportunity as "pain and gain". It often comes as a surprise to NEC users that the terms "pain and gain" do not appear anywhere within the NEC contracts, nor do the terms "target price" or "target cost"!

On completion of the works, the Project Manager makes a preliminary assessment of the Contractor's share by applying the Contractor's share percentage to the difference between the forecast final Price for Work Done to Date and the forecast final total of the Prices. A final assessment is made using the final Price for Work Done to Date and the final total of the Prices.

Whilst NEC practitioners often refer to the terms "pain" and "gain" in Option C and D, these terms do not actually exist within the contract. The Contractor pays or is paid the Contractor's share.

Example

At Completion of the works:

The Contractor's share percentages and the share ranges have been entered into Contract Data Part 1 as follows:

share range	Contractor's share percentage
less than 80%	50%
from 80% to 90%	40%
from 90% to 100%	30%
greater than 100%	50%
The Total of the Prices	= £3,200,000
The Price for Work Done to Date	= £2,400,350
Difference	= £799,650

The share range has been set in increments of 10%

10% of £3,200,000 = £320,000

"From 90% to 100%"

£320,000 × 30% = £96,000

"From 80% to 90%"

£320,000 × 40% = £128,000

"Less than 80%"

£159,650 × 50% = £79,825

Total Contractor's share = £96,000 + £128,000 + £79,825 = £303,825

Question 5.13 How are unfixed materials on or off Site dealt with under the NEC4 contracts?

A common area of confusion is the payment to the Contractor for unfixed materials within or outside the Working Areas, i.e. materials on or off Site. Unlike other forms of contract, the NEC4 has no express provisions for payment for such materials.

Within the Engineering and Construction Contract, for example, in order for the Contractor to be paid for unfixed materials, whether they be on or off Site, he should consider the payment rules for each of the Main Options.

Option A

As the Price for Work Done to Date (PWDD) is based on completed activities, the Contractor should create an activity in the activity schedule for unfixed materials. For example, for a structural steel frame, the Contractor could include four activities:

1   delivery of steel to Site
2   erection of steel to Gridline 1–10
3   erection of steel to Gridline 10–20
4   paint steel.

As each activity is completed, the Contractor is then paid within the assessment following completion of each activity.

Option B

As the PWDD is based on Bills of Quantities, appropriate items may be included as method related charges.

Option C, D, E and F

As the PWDD is the Defined Cost which the Project Manager forecasts will have been paid by the Contractor before the next assessment date plus the Fee, the Contractor must provide accounts and records to show that he has paid for the materials or will have paid for them by the next assessment date.

Note that, in respect of unfixed materials on Site, whatever title the Contractor has to Plant and Materials passes to the Client if it has been brought within the Working Areas and passes back to the Contractor if it is removed from the Working Areas with the Project Manager's permission.

In respect of unfixed materials off Site, whatever title the Contractor has to Plant and Materials passes to the Client if the Supervisor has marked it as for this contract.

Also, in respect of marking, the contract must have identified them for payment, and the Contractor must have prepared them for marking as required by the Scope. This could include setting them aside from other stock, protection, insurance and any vesting requirements.

Question 5.14 How is interest calculated on late payments on an NEC4 Engineering and Construction Contract?

Interest is paid on late payments and is covered by Clause 51.2 to 51.4, interest being paid by the Client if a payment is not made, or the Project Manager does not issue a certificate that it should have issued.

The interest rate is stated in Part 1 of the Contract Data and is assessed on a daily basis from the date the payment should have been made until the date when the late payment is made, calculated using the interest rate in Contract Data Part 1 compounded annually.

The interest due can be calculated on the basis of the following formula:

$$\frac{\text{Payment due} \times \text{interest rate} \times \text{the number of days late}}{365}$$

So, if one assumes the following:

- payment due = £120,000
- payment 7 days late
- interest rate in Contract Data Part 1 = 3%

The calculation is:

$$\frac{£120,000 \times 3\% \times 7\,\text{days}}{365}$$

Interest due = £69.04

Similarly, interest is paid on a correcting amount due to apply to later corrections to certified amounts by the Project Manager and on interest due to a compensation event or as determined by the Adjudicator or the tribunal. The date from which interest should run is the date on which the additional payment would have been certified if there had been no dispute or mistake.

Note that under Option Y(UK)2 (if selected), which relates to the Housing Grants, Construction and Regeneration Act 1996 and the Local Democracy, Economic Development and Construction Act 2009, if payment is late, the Contractor may exercise his right under the Act to suspend performance due to late or non-payment.

It is also a compensation event under Option Y(UK)2 (Clause Y2.5), so the Contractor can recover any Defined Cost plus Fee and/or delay to Completion as a result of exercising his right to suspend all or part of the works.

150 Payment provisions

Question 5.15 The Contractor on an NEC4 Engineering and Construction Contract has failed to submit an application for payment by the assessment date. The Project Manager has stated that no payment is therefore due to the Contractor this month. Is this correct?

Let us start by considering the payment assessment process.

The Project Manager is required to assess the amount due to the Contractor at each assessment date. The first assessment date is decided by the Project Manager to suit the Parties, and will normally be based on the time that the Contractor has been Providing the Works, the Client's procedures and timing for processing and issuing payments, and the Contractor's payment requirements and internal accounting system.

However, the first assessment must be made within the "assessment interval" after the starting date, this is normally inserted in the Contract Data as "four weeks" or "one calendar month".

Clearly, in this respect, some discussion needs to take place between the Project Manager, the Client and the Contractor in order that a mutually agreeable assessment date can be set.

Later assessment dates occur at the end of each assessment interval until

- the Supervisor issues the Defects Certificate
- the Project Manager issues a termination certificate.

Note that there is no provision within the contract for a minimum certificate amount.

The Project Manager assesses the amount due at each assessment date, calculating the Price for Work Done to Date using the rules of the specific Main Option.

The Contractor submits an application for payment to the Project Manager before each assessment date, and must include details of the application, the format to be in accordance with the Scope.

In making his assessment, the Project Manager considers "an application for payment the Contractor has submitted" before the assessment date.

If the Contractor submits an application for payment, the amount due to the Contractor is:

- the Price for Work Done to Date
- plus other amounts to be paid to the Contractor (e.g. Contractor's share, value added tax, etc.)
- less amounts to be paid by or retained from the Contractor (e.g. retention, delay damages).

If the Contractor does not submit an application for payment, the amount due is the lesser of:

- the amount the Project Manager assesses as due at the assessment date, as though the Contractor had submitted an application for payment before the assessment date, and
- the amount due at the previous assessment date.

When you consider these two bullet points, by using the words "the amount due is the lesser of . . .", in effect, the Contractor is *not* due any payment from the Client unless he submits an application for payment. However, there may be an amount due *from* the Contractor *to* the Client.

It is essential that the Contractor either submits a programme with his tender or within the time scale specified within Contract Data Part 1. Failure to do so will entitle the Project Manager to retain one quarter of the Price for Work Done to Date in his assessment of the amount due.

Note that the amount is only withheld if the Contractor has not submitted a programme that shows the information that the contract requires, e.g. method statement, time risk allowances, etc. If the Contractor has submitted a programme that contains all the information that the contract requires, but the Project Manager disagrees with, for example, part of the method statement or the programme has not yet been accepted, then the provision, and the associated amount retained, does not apply.

Question 5.16   We have a contract under the NEC4 Engineering and Construction Contract where the Contractor has stated that he must be paid for compensation events if he has done the relevant work. Is the Project Manager required to certify payments "on account" for compensation events not yet agreed?

The answer to this question is essentially no, as under Clause 66.2 *"when a compensation event is implemented the Prices, the Completion Date and the Key Dates are changed accordingly"*. Once implementation occurs, then the compensation event and its effect is incorporated into the contract.

However, if we take each Main Option in turn:

Option A – Priced contract with activity schedule

The Price for Work Done to Date is defined under Clause 11.2(29):
The Price for Work Done to Date is the total of the Prices for

- each group of completed activities
- each completed activity which is not in a group.

152 Payment provisions

With Option A, if the compensation event has not yet been implemented and compensation events are in the form of changes to the activity schedule (Clause 63.14), there is no activity as yet which, if complete, can be paid.

Option B – Priced contract with bill of quantities

11.2(30) The Price for Work Done to Date is the total of

- the quantity of the work that the Contractor has completed for each item in the Bill of Quantities multiplied by the rate and
- a proportion of each lump sum which is the proportion of the work covered by the item that the Contractor has completed.

In that case, if the compensation event has not yet been implemented and compensation events are in the form of changes to the Bill of Quantities (Clause 63.15), there is no item in the Bill of Quantities as yet against which payment can be made.

Option C – Target contract with activity schedule

11.2(31) The Price for Work Done to Date is the total Defined Cost which the Project Manager forecasts will have been paid by the Contractor before the next assessment date plus the Fee.

With Option C, if the compensation event has not yet been implemented and compensation events are in the form of changes to the activity schedule (Clause 63.14), and the consequent effect on the target, there is no activity as yet against which payment can be made.

However, payment is not based on completed activities, but on Defined Cost, therefore the Contractor can be paid, even though the relevant compensation event has not been implemented.

Option D – Target contract with bill of quantities

With Option D,

11.2(31) The Price for Work Done to Date is the total Defined Cost which the Project Manager forecasts will have been paid by the Contractor before the next assessment date plus the Fee.

With Option D, if the compensation event has not yet been implemented and compensation events are in the form of changes to the Bill of Quantities (Clause 63.15), there is no item in the Bill of Quantities as yet against which payment can be made.

However, payment is not based on the Bill of Quantities but on Defined Cost; therefore the Contractor can be paid, even though the relevant compensation event has not been implemented.

Option E – Cost reimbursable contract

Option F – Management contract

11.2(31) The Price for Work Done to Date is the total Defined Cost that the Project Manager forecasts will have been paid by the Contractor before the next assessment date plus the Fee.

With Option E and F, payment is based on Defined Cost; therefore the Contractor can be paid, even though the relevant compensation event has not been implemented.

Question 5.17 We have heard that there are some substantial changes to the payment terms within the NEC4 Professional Service Contract. What are they?

In order to comprehensively answer this question, let us consider in detail the payment and compensation event provisions under the NEC3 Professional Services Contract and the NEC4 Professional Service Contract.

NEC3 Professional Services Contract

The NEC3 Professional Services Contract required the Consultant to submit staff rates within Contract Data Part 2, identifying the name/designation of the member of staff, and the hourly rate.

Then, dependant on the Main Option chosen, the Consultant was paid a "Time Charge" which is "the sum of the products of each of the staff rates multiplied by the total staff time appropriate to that rate appropriately spent on work in this contract" (see Table 5.1).

It is essential that these staff rates tendered by the Consultant are inclusive of any costs attributable to those members of staff, and must also include profit and any overheads, as there is no other provision for recovering them elsewhere.

The NEC3 Professional Services Contract also provided for expenses to be paid to the Consultant, though this must either be stated by the Employer within Contract Data Part 1 as expenses that the Employer will pay, or by the Consultant within Contract Data Part 2 at the time of tender, as expenses that the Consultant wishes to be paid.

Under Option A, the Consultant was required to prepare forecasts of the total expenses for the whole of the services on a periodic basis and to submit them to the Client (Clause 21.3) including an explanation of the changes since the last forecast.

Under Options C, E and G, the Consultant kept accounts and records of his Time Charge and expenses, and allowed the Client to check them at any time during normal working hours (Clause 52.2). In addition, the Consultant was required to prepare forecasts of the total Time Charge expenses for the whole of

Table 5.1 The Price for Services Provided to Date

Option A	Total of the Prices for each completed activity
Option C	Time Charge for the work which has been completed.
Option E	Time Charge for the work which has been completed.
Option G	Time Charge for the work which has been completed, and a proportion of the lump sum price for each other item on the Task Schedule.

the services on a periodic basis and to submit them to the Client (Clause 21.4), including an explanation of the changes since the last forecast.

Assessing compensation events

The basic rule for assessing compensation events is stated in Clause 63.1.
This is based on the effect of the event upon

- the actual Time Charge for work already done and
- the forecast Time Charge for the work not yet done.

In some circumstances where work has already started but has not been completed, assessment may be based on a combination of the two parts.

The Consultant uses staff rates within Contract Data Part 1 to price his quotation. If he is proposing to use staff for which there is no staff rate, or any equipment or materials for which there is no rate, a proposed rate is used.

NEC4 Professional Service Contract

The NEC4 Professional Service Contract has discontinued the use of staff rates and Time Charge in favour of the use of Defined Cost via a Schedule of Cost Components and a Short Schedule of Cost Components, the same as the Engineering and Construction Contract.

(i) Schedule of Cost Components

The Schedule of Cost Components only applies when Option C or E is used.

Clause 1 – People

This relates to the cost of people who are directly employed by the Consultant, and who are providing a part of the service.

The cost component covers the full cost of employing the people including wages and salaries, contributions, levies or taxes imposed by law, pensions and life insurance, death benefit, occupational accident benefits, medical aid and health insurance, and a vehicle.

It also includes bonuses and incentives, overtime and unsocial hours working, severance, protective clothing, safety training, relocation, medical examinations, passports and visas, travel insurance, including for dependants where appropriate.

Finally it includes for the cost of people not directly employed by the Consultant but who are paid according to time worked on the contract.

Clause 2 – Subcontractors

This component includes for payments to Subcontractors, without taking into account any amounts paid to or retained from the Subcontractor which would result in the Client paying or retaining the amount twice.

Clause 3 – Charges

This component includes a charge for support people and office overhead costs calculated by applying the overhead percentage within the Contract Data to the total of the people items. It includes for provision and use of people, accommodation, equipment, supplies and services required to provide the office and to support people providing the service.

Clause 4 – Insurance

The cost of events for which the Consultant is required to insure and other costs to be paid to the Consultant by insurers are deducted from cost.

(ii) Short Schedule of Cost Components

The Short Schedule of Cost Components is restricted to the assessment of compensation events under Option A but, if the Project Manager and Contractor agree, it may be used for assessing compensation events under Options C and E.

Clause 1 – People

This component is calculated by multiplying each of the People Rates by the total time appropriate to that rate properly spent on the contract.

Clause 2 – Subcontractors

This component covers payments to Subcontractors.

Clause 3 – Insurance

The cost of events for which the Contractor is required to insure and other costs to be paid to the Contractor by insurers are deducted from cost (see Table 5.2).

Note, there is no longer an Option G.

Table 5.2 The Price for Service Provided to Date

Option A	Total of the Prices for each completed activity.
Option C	Total Defined Cost which the Service Manager forecasts will have been paid by the Consultant before the next assessment date plus the Fee.
Option E	Total Defined Cost which the Service Manager forecasts will have been paid by the Consultant before the next assessment date plus the Fee.

Accounts and records

The Consultant is required to prepare forecasts of the total Defined Cost and expenses for the whole of the service, and to keep accounts and records of Defined Cost under Options C and E and, under Option A (if applicable) to prepare forecasts of expenses.

Assessing compensation events

The basic rule for assessing compensation events is stated in Clause 63.1.
This is based on the effect of the event upon

- the actual Defined Cost of the work done by the dividing date
- the forecast Defined Cost of the work not done by the dividing date and
- the resulting Fee.

Chapter 6

Managing compensation events

Notification, pricing and assessing compensation events, assumptions, etc.

Question 6.1 How do the NEC4 contracts deal with unforeseen ground conditions on the Site?

NEC4 Engineering and Construction Contract

In considering this question within NEC4 contracts, let us first consider the Engineering and Construction Contract, by reference to Clause 60.1(12):

> The *Contractor* encounters physical conditions which
>
> - *are within the Site*
> - *are not weather conditions and*
> - *an experienced contractor would have judged at the Contract Date to have such a small chance of occurring that it would have been unreasonable for him to have allowed for them.*
>
> *Only the difference between the physical conditions encountered and those for which it would have been reasonable to have allowed is taken into account in assessing a compensation event.*

Note that the Clause refers to "physical conditions" not exclusively ground conditions, so it could equally relate to conditions above ground.

It is important to interpret this final paragraph correctly because if the Contractor did not allow anything in terms of Price and/or time within his tender for dealing with a physical condition he should only be compensated for the *difference* between what he found and what he *should have allowed*, so the Contractor is unlikely to be compensated for the full value of dealing with the physical condition. If he should have allowed time and/or money within his tender for dealing with a physical condition then this must be considered in assessing a compensation event.

Note that, under Clause 60.2, in judging the physical conditions for a compensation event, the Contractor is assumed to have taken into account:

- the Site Information – this could include site investigations, borehole data, etc.
- publicly available information referred to in the Site Information – this could include reference to public records about the Site
- information obtainable from a visual inspection of the Site – note the use of the term "visual", the Contractor is not assumed to have carried out an intrusive investigation of the Site
- other information which an experienced Contractor could reasonably be expected to have or to obtain – this is a fairly subjective criterion, but precludes the Contractor from relying solely on what is contained within the Site Information. Note the clause refers to "an experienced Contractor" not one who has a detailed local knowledge of the Site.

Under Clause 60.3, "if there is an ambiguity or inconsistency within the Site Information (including the information referred to in it), the Contractor is assumed to have taken into account the physical conditions more favourable to doing the work", this could be the cheaper, easier or quicker alternative.

This complies with the rule of "contra proferentem" (contra = against, proferens = the one bringing forth) in that where a term or part of a contract is ambiguous or inconsistent it is construed strictly against the Party that imposes or relies on it.

It is critical that the Client makes all information in his possession available to the Contractor. He cannot be selective, withholding information with a view to obtaining advantage. It is also important to note that if the Contractor encounters unforeseen physical conditions, which are often but not always ground conditions, he may not necessarily be compensated for the cost and time effect of dealing with it, so he must consider carefully the wording of Clause 60.1(12) and 60.2 to prove his case.

It is also important to recognise that compensation to the Contractor is assessed as the difference between what the Contractor found, and what it would have been reasonable for him to have allowed in his tender, not simply the difference between what he found and what he allowed in his tender.

NEC4 Engineering and Construction Short Contract

The Engineering and Construction Short Contract has the same provisions as the NEC4 Engineering and Construction Contract.

NEC4 Professional Services Contract

There is no provision for unforeseen physical conditions within the Professional Services Contract.

NEC4 Term Service Contract

There is no provision for unforeseen physical conditions within the Term Service Contract.

However, whilst an NEC4 Term Service Contract will not normally include significant excavation work, the possibility of encountering physical conditions such as natural or artificial obstructions should be considered. This risk is possible in say highways or drainage maintenance contracts.

Also, as stated above, Clause 60.1(12) within the NEC4 Engineering and Construction Contract refers to "physical conditions" not exclusively ground conditions, so it could equally relate to conditions above ground that could affect an NEC4 Term Service Contract.

It is therefore suggested that, dependent on the Scope, an additional compensation event should be considered for the NEC4 Term Service Contract, similar to that in the NEC4 Engineering and Construction Contract.

Question 6.2 How do the NEC4 contracts deal with delays and/or costs incurred by exceptionally adverse weather conditions?

The NEC4 contracts deal with weather in different ways:

Engineering and Construction Contract

Under the Engineering and Construction Contract, we must consider Clause 60.1(13):

A weather measurement is recorded

- *within a calendar month,*
- *before the Completion Date for the whole of the works and*
- *at the place stated in the Contract Data*

the value of which, by comparison with the weather data, is shown to occur on average less frequently than once in ten years.

Only the difference between the weather measurement and the weather which the weather data show to occur on average less frequently than once in ten years is taken into account in assessing a compensation event.

With regard to weather related delays, most contracts use the words "exceptionally adverse weather" or "exceptionally adverse climatic conditions" leaving it to the Parties to determine, and in many cases to argue what is meant by the words "exceptionally adverse". "Exceptionally" may be defined as "more or less than normal" whilst adverse may be defined as "unfavourable" in the sense that it impacted upon the Contractor's progress, which has to take into account the weather that would reasonably have been expected, bearing in mind the location of the site, the time of year and what the Contractor was intending to do at the time the weather condition occurred.

The ECC provides a more objective approach by referring and comparing to weather data provided by an independent Party such as a meteorological office, an airport or military base.

Only the difference between the weather measurement and the weather that the weather data show to occur less frequently than once in ten years is taken into account in assessing a compensation event, so the Contractor will not be compensated for the full impact of the weather event as weather likely to occur within a ten-year period is the Contractor's risk, and it is assumed that the Contractor has already allowed for that in terms of price and programme.

Contract Data Part 1 defines the place where the weather is to be recorded. Note that Clause 60.1(13) states "at the place stated in the Contract Data" so it is vital that the historic weather data *and* the current weather measurements are recorded at the same place. It is important that the place chosen will truly reflect the likely conditions that will be encountered at the Site.

It also lists weather measurements for each calendar month in respect of:

- *the cumulative rainfall (mm)*
- *the number of days with rainfall more than 5mm – whilst location and time of year is clearly a factor, from analysis of meteorological records, 5mm rainfall in a day is a fairly low level*
- *the number of days in the month with minimum air temperature less than 0 degrees Celsius, and*
- *the number of days in the month with snow lying at a stated time GMT – there is no measure of how much snow, merely that it is lying, presumably on the ground, at a stated time.*

There is also provision for adding other measurements which could include wind speed, and other weather related data. This might apply at, say, coastal or mountain locations.

Note, also, that the weather measurement is recorded before the Completion Date for the whole of the works and at the place stated in the Contract Data.

As a compensation event under Clause 60.1(13) occurs when a weather measurement is recorded within a calendar month, the value of which, by comparison with the weather data, is shown to occur on average less frequently than once in ten years, there is a "trigger point" in the month at which the weather, for example the cumulative rainfall exceeds the "once in ten years" test. Note that the compensation event occurs once the trigger point has been exceeded, not once it has an effect on the progress of the works (see Figure 6.1).

The weather up to the trigger point is at the Contractor's risk in terms of programme and cost and is deemed to have been included within the Contractor's

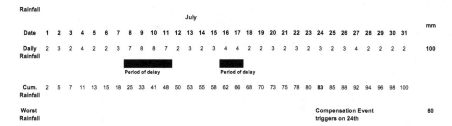

Figure 6.1 Weather related compensation events

tender, but once the trigger point is reached, there are two schools of thought as to how the compensation event operates:

1   If the trigger point is reached on say, the 24th of the month, one must consider the *whole* month, but only the difference between the weather measurement and the weather which the weather data show to occur on average less frequently than once in ten years is taken into account in assessing a compensation event.
2   If the trigger point is reached on say, the 24th of the month, only delays incurred by the weather *after* the 24th are compensated for.

In either case, one must consider the worst weather in ten years, not the average weather. Also, once a compensation event exists it is incumbent upon the Contractor to submit a quotation showing any effect that the weather had on the Prices and/or any delay to Completion and/or Key Dates. There is no automatic entitlement to be paid for the weather, just because the trigger point was exceeded.

It has to be said that assessing the effect of weather on an objective basis, rather than the traditional subjective basis, is essential, particularly as the NEC4 Engineering and Construction Contract awards money as well as time, whilst most non NEC contracts only award time, but the NEC approach presents its own challenges!

Professional Service Contract

Unlike the NEC4 Engineering and Construction Contract and Engineering and Construction Short Contract, the Professional Service Contract does not include a compensation event for delays and/or costs caused by adverse weather.

While this may at first not seem to be an important issue for the Consultant, for example, in a drawing office designing a project rather than the Contractor on site building it, it can be a concern and a major risk for a Consultant carrying out other professional services that may be affected by the weather, for example, site inspections, surveys, etc.

Where external work is involved, it may be appropriate for the Client to add a Z clause for weather-related risks similar to Clause 60.1(13) of the Engineering and Construction Contract as above.

Term Service Contract

There is no provision for delays and/or costs incurred by adverse weather conditions under the Term Service Contract.

Again, where external work is involved, it may be appropriate for the Client to add a Z clause for weather-related risks similar to Clause 60.1(13) of the Engineering and Construction Contract.

Question 6.3 Do the NEC4 contracts include provision for "force majeure" events?

No, the NEC4 contracts do not specifically provide for "force majeure" events in the sense of describing them, and defining what is a force majeure event, though, if we consider the Engineering and Construction Contract, this includes Clause 60.1(19), which many could define as a "force majeure type of event".

The clause states:

> An event which
>
> - stops the Contractor completing the whole of the works or
> - stops the Contractor completing the whole of the works by the date shown on the Accepted Programme,
>
> and which
>
> - neither Party could prevent
> - an experienced contractor would have judged at the Contract Date to have such a small chance of occurring that it would have been unreason- able for him to have allowed for it and
> - is not one of the other compensation events stated in the contract.

Clause 60.1(19) is a new "no fault" clause first introduced under NEC3, and dealing with an event neither Party could prevent, which stops the Contractor completing the works.

It is also tied in to Clause 19.1 "Prevention" following which, if the event occurs, the Project Manager instructs the Contractor how to deal with the event.

There has been much confusion and debate regarding these clauses since NEC3 was first published in 2005.

The intention of the drafters appears to be that this would be a force majeure (superior force) clause, often referred to as an "Act of God" in other contracts and legal documents, i.e. significant events such as flooding, earthquakes, volcanic eruptions and the associated dust clouds and other natural disasters that prevent the contracting Parties from fulfilling their obligations under the contract, the clause essentially freeing them, or giving some relief from their liabilities. Whilst this appears to be what the drafters had intended, on closer examination the application of this clause is far wider. If you take each element in turn:

The event

- *stops the Contractor completing the whole of the works or*
  Due to the event, the *whole of the works* was never completed.
- *stops the Contractor completing the whole of the works by the date shown on the Accepted Programme*
  Due to the event, completion of the *whole of the works* was delayed.
- *neither Party could prevent*
  Clearly, if either Party could have prevented the event then they would, or should have. The clause refers to preventing the event, rather than taking some mitigating action to lessen its effect.
- *an experienced contractor would have judged at the Contract Date to have such a small chance of occurring that it would have been unreasonable for him to have allowed for it*
  At tender stage, the likelihood of the event was so low, the Contractor would not have allowed for it.
- *is not one of the other compensation events stated in this contract*
  If the matter could have been dealt with as one of the other compensation events then it should be. 60.1(19) then clearly deals with the exceptional event.

Let's look at an example of where this clause would at least cause confusion.

Clause 60.1(13) has been deleted from the contract by means of a Z clause; one would then assume that the risk of unforeseen weather conditions is the Contractor's and he should have allowed for it. The site is affected by a severe storm which occurs less frequently than once in 50 years, which delays completion of the works by one week.

The Contractor states that:

- *The weather stopped him completing the works by the date shown on the Accepted Programme.*
  Assume that the Contractor can prove that.
- *Neither Party could prevent the event.*
  Whilst one can take measures to lessen the impact of a storm, one cannot prevent the storm from occurring.

- An experienced contractor would have judged at the Contract Date to have such a small chance of it occurring that it would have been unreasonable to have allowed for it.

  The Contractor states that he could not have predicted or allowed for such a rare event and its consequences.
- It is not one of the other compensation events.

  If Clause 60.1(13) has been deleted from the contract by means of a Z clause, one could argue that it is no longer one of the other compensation events.

Is the Contractor then entitled to a compensation event for the weather?

Could Clause 60.1(19) also extend to loss or damage occasioned by insurable risks such as fire, lightning, explosion, storm, flood, earthquake, terrorism?

Although both Schedules of Cost Components require that the cost of events for which the contract requires the Contractor to insure should be deducted from cost, the Contractor could still recover any delay to Completion through the compensation event.

If Clause 60.1(19) is intended as a form of force majeure provision, rather than including a fairly long winded definition within the clause, why not just state: "A force majeure event occurs" and allow the Parties to interpret this internationally recognised term as they do in other contracts.

Question 6.4   How long do the Project Manager and the Contractor have on an NEC4 Engineering and Construction Contract to notify compensation events?

Compensation events may either be notified by the Project Manager, or the Contractor can notify a compensation event to the Project Manager.

1. The Project Manager notifies the Contractor of the compensation event at the time of giving the instruction, issuing a certificate, or changing an earlier decision (Clause 61.1). It also instructs the Contractor to submit quotations, unless the event arises from a fault of the Contractor or the event has no effect upon Defined Cost, Completion or meeting a Key Date. The Contractor puts the instruction or changed decision into effect.
2. The Contractor can notify a compensation event, but must do so within eight weeks of becoming aware of the event, otherwise it is not entitled to a change in the Prices, the Completion Date or a Key Date, unless the event arises from the Project Manager or the Supervisor giving the instruction, issuing a certificate, or changing an earlier decision. In other words, the Project Manager should have notified the event to the Contractor but did not (Clause 61.3).

   The intention of the clause is to compel the Contractor to notify compensation events promptly, otherwise any entitlement to additional time and money is lost. Note that the eight week rule requires the Contractor to notify the compensation event to the Project Manager within eight weeks of becoming aware of it, not just to have given early warning or mentioned the possibility of a compensation event in a discussion with the Project Manager.

The clause therefore covers compensation events initiated by the Contractor, rather than the Project Manager, examples being the weather, unforeseen ground conditions, which the Project Manager may not be aware of unless the Contractor had notified it.

An example of the Project Manager failing to notify a compensation event when it should have would be if an instruction was issued by the Project Manager to the Contractor changing the Scope, but at the time the Project Manager did not notify the compensation event, and the Contractor in turn did not notify either. Even if the compensation event is not notified within eight weeks, the responsibility remains with the Project Manager as it gave the instruction and should have notified the event to the Contractor but did not.

Various opinions have been published about the enforceability and effectiveness of time bars in contracts such as NEC, commentators particularly debating whether the clause is a condition precedent to the Contractor being able to recover time and money, and whether a Party, the Client can benefit from its own breach of contract to the detriment of the injured Party, the Contractor.

For example, if the Client does not provide something which it is to provide by the date for providing it shown on the Accepted Programme, this is a valid compensation event under Clause 60.1(3), but can the Client prevent the Contractor from receiving any remedy because the Contractor failed to notify the Project Manager within the eight week limit, and possibly, if Completion is delayed the Contractor will have to pay delay damages?

The author, whilst not being a lawyer, is of the opinion that the Parties are clear as to what the terms of their agreement are at the time the contract is formed. It is also clear what happens if the Contractor does not notify a compensation event within the time stated within Clause 61.3; it is protected against notifications which the Project Manager should have given but did not, and therefore the time bar must be effective and enforceable.

As there are likely to be several compensation events during the period of the contract, a schedule of compensation events should be kept, identifying and numbering each, with additional information about each one.

The author has found the schedule useful for tracking compensation events as the project progresses, but also at the end of each project to review the schedule and each compensation event notified during the life of the project as "lessons learned" to consider in drafting Scopes for future contracts.

Clause 61.4 covers three possible outcomes to the Contractor's notification of a compensation event under Clause 61.3.

Negative reply

If the Project Manager responds by stating that an event notified by the Contractor

- *arises from a fault of the Contractor*
- *has not happened and is not expected to happen*
- *has not notified within the timescales set out in these conditions of contract*
- *has no effect upon Defined Cost, Completion or meeting a Key Date or*
- *is not one of the compensation events stated in the contract*

the Project Manager notifies the Contractor of its decision that the Prices, Completion Date and the Key Dates are not to be changed and states the reason. Note that the Project Manager only needs to name *one* of these as its reason for refusing a compensation event.

Positive reply

If the Project Manager decides otherwise, it notifies the Contractor that it is a compensation event, and instructs him to submit quotations.

No reply

If the Project Manager does not reply within the time allowed, then the Contractor may notify the Project Manager to that effect. If the Project Manager does not reply within two weeks of the Contractor's notification, then it is deemed acceptance that the event is a compensation event and an instruction to submit quotations. It is a condition precedent upon the Contractor that the notification be given before deemed acceptance can occur.

Question 6.5 We are working on an Option A (priced contract with activity schedule) under the NEC4 Engineering and Construction Contract and have submitted quotations for compensation events to the Project Manager, but he has not responded, and he has said he will discuss them with us when we submit the Final Account. Can he do this?

First, there is no such thing as a Final Account under any of the NEC4 contracts – it is not a term that has ever been used!

The NEC4 Engineering and Construction Contract does not have the equivalent of a Final Account or Final Certificate which is found in other contracts, certifying that the contract has fully and finally been complied with, and that issues such as payments, compensation events and defects have all been dealt with and, effectively, the contract can be closed.

Whilst some may say that, because of the compensation event provisions there is no Final Account, and the Defects Certificate confirms correction of any outstanding defects, there is no need for a Final Certificate but some questions could still remain such as when, under Option E, is it too late for the Contractor to submit a cost?

However, NEC4 has introduced a new provision (Clause 53) where, in the case of the Engineering and Construction Contract, the Project Manager makes a final assessment of amounts due to the Contractor, in effect giving closure, at least to financial aspects of the contract.

The Project Manager makes an assessment of the final amount due to the Contractor and certifies a payment, no later than:

- four weeks after the Supervisor issues the Defects Certificate or
- thirteen weeks after the Project Manager issues a termination certificate.

Similar to a normal payment, the Project Manager gives the Contractor details of the assessment and payment (by either Party) is made within three weeks of the assessment date or, if a different period is stated in the Contract Data, within the period stated.

If the Project Manager does not make the final assessment within the time allowed, the Contractor may issue to the Client an assessment of the final amount due. If the Client agrees with the assessment, a final payment is issued within three weeks of the assessment date or, if a different period is stated in the Contract Data, within the period stated.

The assessment of the final amount due is conclusive, unless
If Option W1 is selected, a Party:

- refers a dispute about the assessment of the final amount due to the Senior Representatives within four weeks of the assessment being issued
- refers any issues not agreed by the Senior Representatives to the Adjudicator within three weeks of the list of issues not agreed being produced, or when it should have been issued, and
- refers to the tribunal its dissatisfaction with a decision of the Adjudicator regarding the final amount due within four weeks of the decision being made.

If Option W2 is selected, a Party:

- refers a dispute about the assessment of the final amount due to the Senior Representatives or to the Adjudicator within four weeks of the assessment being issued, but it may omit this stage by virtue of W2.2(1)
- refers any issues not agreed by the Senior Representatives to the Adjudicator within three weeks of the list of issues not agreed being produced, or when it should have been issued, and
- refers to the tribunal its dissatisfaction with a decision of the Adjudicator regarding the final amount due within four weeks of the decision being made.

If the above applies under Options W1 or W2, the assessment of the final amount due is changed to include:

168 Managing compensation events

- any agreement the Parties reach, and
- a decision of the Adjudicator which has not been referred to the tribunal within four weeks of the decision.

If Option W3 is selected, a Party:

- refers a dispute about the assessment of the final amount due to the Dispute Avoidance Board
- refers to the tribunal its dissatisfaction with a recommendation of the Dispute Avoidance Board within four weeks of the recommendation being made.

The assessment of the final amount due is changed to include:

Any agreement the Parties reach and a decision of the Adjudicator or recommendation of the Dispute Avoidance Board which has not been referred to the tribunal within four weeks of the decision or recommendation.

A changed assessment becomes conclusive evidence of the final amount due.

Question 6.6   How should a Contractor under the NEC4 Engineering and Construction Contract prepare a quotation for a compensation event?

Quotations for compensation events are based on their effect on changes to the Prices using Defined Cost plus Fee, and any delay to the Completion Date and any Key Dates. This is different from most standard forms of contract where variations are valued using the rates and prices in the contract as a basis.

The reason for this policy within the Engineering and Construction Contract is that no compensation event that is the subject of a quotation is due to the fault of the Contractor, or relates to a matter that is at his risk under the contract. It is therefore appropriate to reimburse the Contractor his forecast additional costs or Defined additional costs arising from the compensation event.

Compensation events are assessed as the effect of the compensation event upon:

- the actual Defined Cost of work done by the dividing date
- the forecast Defined Cost of the work not done by the dividing date, and
- the resulting Fee.

It is worth some clarification here regarding the term "dividing date".

Clause 63.1 in previous editions of the ECC stated:

- the actual Defined Cost of the work already done
- the forecast Defined Cost of the work not yet done, and
- the resulting Fee.

This led to some confusion as to when, in preparing his quotation, the Contractor changed from "actual Defined Cost" to "forecast Defined Cost", so the new term "dividing date" was included within the NEC4 ECC.

Again, as previously stated above, for a compensation event arising from the Project Manager or Supervisor giving an instruction or notification, issuing a certificate or changing an earlier decision, the "dividing date" is the date of that communication. For other compensation events, the "dividing date" is the date of the notification of the compensation event.

So, in preparing his quotation, the Contractor uses actual Defined Cost of work which has been done by the dividing date, and forecast Defined Cost of the work yet to be done by the dividing date.

Under Options A to D, if the Project Manager and Contractor agree, rates and lump sums may be used instead of Defined Cost.

Note that such rates and prices do not have to be from the Activity Schedule (Options A and C) or Bill of Quantities (Options B and D), thereby allowing a mutually agreed fair rate or price to be agreed.

The Contractor must ensure that he includes within his quotation for cost and time which have a significant chance of occurring and are his risk (Clause 63.8). This should include for adverse weather conditions and physical conditions which would not be compensation events.

Changes to the Prices take the form of changes to the Activity Schedule (Options A and C) or changes to the Bill of Quantities (Options B or D).

Under Clause 63.5, a delay to the Completion Date is assessed as the length of time that, due to the compensation event, planned Completion is later than planned Completion as shown on the Accepted Programme current at the dividing date.

Similarly, a delay to a Key Date is assessed as the length of time that, due to the compensation event, the planned date when the Condition stated for a Key Date will be met is later than that shown on the Accepted Programme current at the dividing date.

When assessing delay only those operations that the Contractor has not completed and that are affected by the compensation event are changed. Clearly, this emphasises the need for an Accepted Programme to be in place in order that it can be used to make correct assessments.

No compensation event can result in a reduction in the time for carrying out the works, i.e. an earlier Completion Date. Only acceleration as agreed under Clause 36 can result in an earlier Completion Date.

Any time risk allowances that the Contractor has allowed are preserved by this clause, as assessment of the compensation event is based on entitlement rather than need. Allowances for risk must be included in forecasts of Defined Cost and Completion in the same way that the Contractor allows for risks when pricing his tender. Float within the Accepted Programme is, however, available to mitigate or avoid any consequential delay to the Completion Date.

Question 6.7 **We are Contractors carrying out a refurbishment contract under NEC4 Engineering and Construction Contract Option A (priced contract with activity schedule) and have had several small compensation events, each with time delays of less than one day. However, the total cumulative effect of all these compensation events will cause a delay to the Completion Date of approximately nine days. How do we recover these delays and their associated costs under the contract?**

Under the NEC4 Engineering and Construction Contract (Clause 62.2), the Contractor must assess and include within his quotation for a compensation event for any changes to the Prices, any delay to the Completion Date and any Key Dates, and submit details of his assessment within his quotation. In assessing any delay to the Completion Date the delay is the length of time that, due to the compensation event, planned Completion is later than planned Completion as shown in the Accepted Programme.

This is fine where the Contractor forecasts that a compensation event will impact on the Completion Date by say two weeks; the delay and any associated effect on Defined Cost plus Fee will then be included within the quotation.

The problem is that, often with refurbishment projects, a compensation event may arise as a result of a Project Manager's instruction to carry out a minor piece of work, for example to remove and replace the skirtings to part of a room instead of retaining and repainting the old skirtings.

In this case, the Contractor will assess the effect on his Defined Cost (plus Fee percentage) including removing the old skirtings, making good the wall, replacing with new skirtings and painting, but although it may in theory add a couple of hours to the programme and in turn the completion of the project, the Contractor would probably not include anything for delays to the Completion Date and any Key Dates within his quotation, and in fact that compensation event on its own would probably not cause any delay to the Completion Date or any Key Dates anyway, so he could not include for the delay and the associated costs.

However, if this happens on several occasions, which again could be likely on a refurbishment project, then the cumulative delay of several compensation events could soon become apparent.

Whilst the author is firmly of the belief that Parties should fully comply with the contract, sometimes you have to apply a practical and common sense solution to its compliance.

With situations like this it is probably best to mutually agree between the Contractor and the Project Manager that compensation events may be "batched together", maybe in terms of "changes to scope of work for skirtings", or just collected on a fortnightly or monthly basis, to give a collective and cumulative effect on Defined Cost plus Fee, the Completion Date and any Key Dates.

In order to maintain some control, the Project Manager should state assumptions about the event in his instruction to the Contractor to submit each quotation, e.g. "do not include for any delay to the Completion Date", then notify a correction to an assumption under Clause 61.6 and then deal with it under Clause 60.1(17) as a compensation event.

Question 6.8   We have a Consultant carrying out survey work under Option A (priced contract with activity schedule) of the NEC4 Professional Service Contract. The survey has been substantially delayed due to extremely bad weather conditions. Is such a delay the Consultant's own risk?

As has been stated previously, the Professional Service Contract has no equivalent of the Engineering and Construction Contract Clause 60.1(13) which gives a compensation event for a weather related event, so one would then assume that the risk of unforeseen weather conditions in this case is the Consultant's and he should have allowed for it.

But let us consider Clause 60.1(12) which could, at least in theory, change this. Clause 60.1(12) states:

An event which

- *stops the Consultant completing the whole of the service, or*
- *stops the Consultant completing the whole of the service by the date for planned Completion shown on the Accepted Programme*

and which

- *neither Party could prevent*
- *an experienced consultant would have judged at the Contract Date to have such a small chance of occurring that it would have been unreasonable to have allowed for it, and*
- *is not one of the other compensation events stated in the contract.*

Example

The Consultant's progress with an external survey is affected by a very severe storm which occurs less frequently than once in 50 years, which delays completion of the works by 1 week.

The Consultant states that:

- the weather stopped him completing the survey, and in turn the whole of the service, by the date shown on the Accepted Programme.
- it can prove that neither Party could prevent the event. While one can take measures to lessen the impact of a storm, one cannot prevent the storm from occurring.

(continued)

172 Managing compensation events

> *(continued)*
> - it could not have predicted or allowed for such a rare event and its consequences.
> - it is not one of the other compensation events. There is no weather related compensation event in the contract, but has Clause 60.1(12) now created one?
>
> Could the Consultant then be entitled to a compensation event for the unforeseen weather conditions under Clause 60.1(12)?

Question 6.9   We have a compensation event in an NEC4 Engineering and Construction Contract where, in discussion with the Contractor during an early warning meeting he has stated that it is extremely difficult to price one element of the work, and we agree with him. The Contractor has suggested including a Provisional Sum within his quotation, which he can adjust later. Can he do this?

First, it is essential to note that the NEC contracts do *not*, and have never provided for Provisional Sums as used in other contracts, where there are elements of work which are not designed or cannot be sufficiently defined at tender stage, and therefore a sum of money is included by the Client in the Bill of Quantities or other pricing document to cover the item.

When the Contractor is instructed to submit a quotation, there may be a part of the quotation that is too uncertain to be forecast reasonably. In this case, the Project Manager should state what the Contractor should assume, which could be a cost and/or time effect (Clause 61.6).

Subsequently, when the effects are known or it is possible to forecast reasonably it notifies a correction to the assumption and it is dealt with as a correction under Clause 60.1(17).

Note: The Project Manager must state the assumption, if the Contractor makes assumptions when pricing the compensation event, then they are not corrected.

> **Example**
>
> The Contractor is instructed by the Project Manager to excavate a trial pit to establish the nature of the sub strata in a part of the Site, so that the design for a new foundation for an oil storage tank can be finalised.
>
> The Project Manager tells the Contractor that it and the structural engineer will inspect the sides to the excavation as the Contractor progresses with the work and will instruct the Contractor when to stop excavating.

> As the plan size and depth of the excavation is uncertain, the Contractor is unable to make a reasonable forecast of proposed changes to the Prices and any delay to Completion, so the Project Manager states that the Contractor should allow an excavation 4 metres × 4 metres and to a depth of 3 metres, and to backfill the excavation on completion when instructed. The Contractor submits its quotation which is accepted by the Project Manager, the compensation event is then implemented.
>
> Subsequently, the excavation is actually required to be 4 metres × 4 metres and to a depth of 4.5 metres.
>
> The Project Manager therefore notifies a correction to an assumption, which is a compensation event under Clause 60.1(17) and instructs the Contractor to submit a quotation.
>
> The resulting quotation forecasts the Defined Cost of the original trial pit, forecasts the Defined Cost of the actual trial pit and the fee percentage is then added to the difference.

Note that no compensation event is notified after the issue of the Defects Certificate.

Question 6.10 How should a Project Manager on an NEC4 Engineering and Construction Contract make his own assessment of a compensation event?

Under Clause 64.1, the Project Manager may, for the following reasons, assess a compensation event:

- *if the Contractor has not submitted the quotation and details of its assessment within the time allowed*
- *if the Project Manager decides that the Contractor has not assessed the compensation event correctly in the quotation and it does not instruct the Contractor to submit a revised quotation*
- *if, when the Contractor submits quotations for a compensation event, the Contractor has not submitted a programme or alterations to a programme which the contract requires him to submit, or*
- *if, when the Contractor submits quotations for a compensation event, the Project Manager has not accepted the Contractor's latest programme for one of the reasons stated in the Contract.*

These are all derived from some failure of the Contractor either to submit a quotation, to assess the compensation event correctly or to submit an acceptable programme.

If the Project Manager makes his own assessment he should put itself in the position of the Contractor, giving a properly reasoned assessment of the effect of the compensation event, detailing the basis of his calculations and providing the Contractor with details of that assessment.

A Project Manager's assessment is not simply the quotation returned to the Contractor with "red pen" reductions down to a figure the Project Manager is prepared to accept.

It is also important to recognise that the Project Manager is not sending his assessment to the Contractor for his acceptance; it is the final decision under the contract, the Contractor's only remedy being adjudication.

Under Clause 64.2, the Project Manager assesses the programme for the remaining work and uses it in the assessment of a compensation event if:

- there is no Accepted Programme
- the Contractor has not submitted a programme or alterations to a programme as required by the contract, or
- the Project Manager has not accepted the Contractor's latest programme for one of the reasons stated in the contract.

The Project Manager notifies the Contractor of his assessment and gives details within the period allowed for the Contractor's submission of his quotation for the same compensation event.

Under Clause 64.4, if the Project Manager does not assess a compensation event within the time allowed, the Contractor may notify him to that effect. If the Project Manager does not reply to the notification within two weeks, the Contractor's notification is treated as acceptance of the quotation by the Project Manager.

The only remedy the Project Manager has, if he later finds the quotation is not acceptable, is to refer it to adjudication. However, it must be remembered that the role of the Adjudicator and the adjudication process is to enforce the contract, therefore if the Contractor's quotation has been submitted in accordance with the contract, then the Client will probably be unsuccessful in the adjudication.

Question 6.11 The Project Manager on an NEC4 Engineering and Construction Contract has instructed the omission of a part of the works. Can the Contractor claim for loss of profit on the omitted works?

When the Project Manager gives an instruction changing the Scope by omitting work, the method of changing the Prices and dealing with the Completion Date is often misunderstood.

The tendency is for Parties to simply delete or even to just ignore the item in the pricing document (activity schedule or bill of quantities); for example, in an Option A contract, the activity in the activity schedule which represents the omitted work is simply reduced or not paid, or in an Option B contract the quantity in the bill of quantities is reduced, or if a single item in the bill, the item is just not paid. Admittedly, this is usually the correct way to price change in a non NEC contract, but in an NEC contract, both assumptions are incorrect.

Omitted work due to a change to the Scope is assessed not by omitting or remeasuring the item in the relevant pricing document, but by forecasting the Defined Cost of that omitted work and adding the fee percentage. The resulting amount is then adjusted against the value in the pricing document.

It must be remembered that the principle with compensation events in terms of their financial value is that the Contractor *neither gains nor loses* as a result of the compensation event; the Contractor is compensated so that he is in the *same* financial position *after* the event as he would have been *before* the event.

If a Contractor has included low rates in his tender for work that is subsequently omitted it is quite probable that the application of forecast defined Cost plus Fee will give rise to a negative value; this is testament to the fact that the Contractor would have made a loss if the work had *not* been omitted. Similarly, if high rates exist, the Contractor will retain the margin he would have made if he had carried out the work.

Example

In an Option A contract, the Activity Schedule includes an activity entitled "boundary fence", the activity having a price of £26,000. The Project Manager gives an instruction to omit the boundary fence; this is a change to the Scope and therefore a compensation event under Clause 60.1(1).

The tendency is to assume that the Contractor is simply not paid for that activity as it will not be carried out; however, the correct way to assess the change to the Prices is to forecast the Defined Cost of the work not yet done and add the resulting Fee.

Let us assume that the Contractor has already placed an order in the sum of £20,100 with a fencing subcontractor, and can therefore prove what its costs would have been had the work not been omitted and the fee percentage is 8%.

The Forecast Defined Cost is then the value of the subcontract order:

Subcontract Order to supply and install fencing	£20,100.00
Add 8%	£1,608.00
Change to the Prices	£21,708.00

The amount of the quotation, assuming it has been accepted by the Project Manager, is then used to change the Prices, so either a new activity is then

(continued)

176   Managing compensation events

> *(continued)*
>
> inserted to the value of −£21,708.00 or the activity priced at £26,000.00 is omitted and replaced with an activity priced at £26,000.00 − £21,708.00 = £4,292.00.
>
> The Contractor in this case retains the profit he would have made if the boundary fence had not been omitted. Clearly, if the change to the Prices had been more than £26,000 the Contractor would retain the loss.

The same principle would apply with all the main options; for example, if it was an Option B contract and the boundary fence was an item in the Bill of Quantities, then the Bill of Quantities would again be changed in the same way, and again the Contractor retains the profit (or loss) it would have made if the boundary fence had not been omitted.

Question 6.12   We have an Engineering and Construction Contract Option B (priced contract with bill of quantities), and a description of a product in the Bill of Quantities states "or similar approved". If the Contractor then proposes a cheaper alternative (which is approved), does this saving get administered as a negative compensation event?

The phrase "or similar approved" must be stated in the Scope, not just the bill of quantities, as it is the Scope that basically tells the Contractor what to do in order to Provide the Works.

Assuming this is the case then, providing there has been such an approval, the Contractor has complied with the Scope. There is no instruction needed to change it, as nothing has been changed, and there is no compensation event, negative or otherwise.

Note that there is no such thing as a "negative compensation event", though there may be a compensation event where the change to the Prices is a minus figure.

The Scope was written to allow this, providing it was approved, and that is what has happened. The financial effect of that depends on what Main Option you are using.

In Main Options A (priced contract with activity schedule) and B the Contractor keeps all the benefit (or possibly the burden).

However, in Main Options C and D (target contracts with bill of quantities and activity schedule respectively) the benefit (or burden) is shared.

Note also that acceptance rather than approval is generally the language used in NEC contracts.

Question 6.13   We would like to include value engineering within the NEC4 contracts and to reward the Contractor if he proposes a change

that can save the Client money. Is there any provision for value engineering within the NEC4 contracts?

Many users of the previous NEC3 contracts do not realise that there has always been provision for value engineering within the contracts, for example within the Engineering and Construction Contract Option C (Target Contract with Activity Schedule) and Option D (Target Contract with Bill of Quantities) there is Clause 63.11.

If the Employer changes the Works Information, perhaps because his design team have carried out a value engineering exercise to generate savings and he wishes to incorporate their findings within the contract, the Project Manager gives an instruction changing the Works Information and it is a compensation event (Clause 60.1(1)), the Contractor then submits a quotation for the saving and the Prices are reduced (the target is lowered) accordingly, then the cost savings will be accrued through the Defined Cost paid to the Contractor.

However, the Contractor may have carried out a value engineering exercise and submitted proposals to the Project Manager for his (and the Employer's) consideration. If the Project Manager then accepts the proposals and the Works Information is changed accordingly, then the first bullet of Clause 63.11 is applicable, i.e.:

> *If the effect of a compensation event is to reduce the total Defined Cost and the event is*
>
> - *a change to the Works Information, other than a change to the Works Information provided by the Employer which the Contractor proposed and the Project Manager has accepted . . .*
>
> *the Prices are reduced.*

This clause has to be read a couple of times as it seems quite convoluted but what it means is that the Prices are reduced (the target is lowered) *unless* there is a change to the Works Information provided by the Employer which the Contractor proposed and the Project Manager has accepted, in which case the Prices are *not* reduced and the cost savings accrued through the Defined Cost paid to the Contractor will lead to a sharing of the benefit between the Employer and the Contractor.

Clause 63.13 of the NEC4 Engineering and Construction Contract has replaced Clause 63.11 (above), whilst Clause 63.12 (Options A and B) states:

> *If the effect of a compensation event is to reduce the total Defined Cost and the event is a change to the Scope, other than a change to the Scope provided by the Client, which the Contractor proposed and the Project Manager accepted, the Prices are reduced by an amount calculated by multiplying the assessed effect of the compensation event by the value engineering percentage (In Contract Data Part 1).*

There are also two new further provisions incorporated within the NEC4 Engineering and Construction Contract:

1   Contractor's proposals (Clause 16)

Clause 16 is a new provision, where the Contractor may propose to the Project Manager that the Scope is changed to reduce the amount the Client pays to the Contractor. This could also be considered as part of a value engineering or a risk management process.

An example could be that a material specified within the Scope could be substituted with another, less expensive material that is equally effective, or a specified methodology could be substituted with a more rational, and again less expensive methodology, for example the use of mobile access platforms rather than a full static scaffold.

Within four weeks of making the proposal, the Project Manager may:

- *accept the proposal and issue an instruction changing the Scope*
- *inform the Contractor that the Client is considering the proposal and instruct the Contractor to submit a quotation, or*
- *inform the Contractor that the proposal is not accepted.*

2   Option X21 (Whole life cost)

X21 is a new Secondary Option, not formerly included within the NEC3 contract.

Under this Option, the Contractor may propose to the Project Manager that the Scope is changed in order to reduce the cost of operating and maintaining the Affected Property.

If the Project Manager is prepared to consider the change, the Contractor submits a quotation to the Project Manager which includes:

- *a detailed description*
- *the forecast cost reduction to the Client of the asset over its whole life*
- *an analysis of the resulting risks to the Client*
- *the proposed change to the Prices and*
- *a revised programme showing any changes to the Completion Date and Key Dates.*

The Project Manager consults with the Contractor about the quotation and replies within the period for reply either accepting or not accepting the quotation.

If the quotation is accepted, the Project Manager changes the Scope, the Prices, the Completion Date and any Key Dates and accepts the revised programme. The change to the Scope is not a compensation event.

Question 6.14 What does the term "implemented" mean within the NEC4 contracts when referring to compensation events?

The word "implemented" has a specific meaning under NEC4 contracts.

Under Clause 66.1 of the NEC4 Engineering and Construction Contract, the Project Manager implements each compensation event by notifying the Contractor of his acceptance of the Contractor's quotation, notifying the Contractor of his own assessment, or a Contractor's quotation may be treated as accepted subject to Clause 62.6 or 64.4.

Under Clause 66.3, the assessment of an implemented compensation event is not revised except as stated in the conditions of contract.

If the subsequent records of resources on work actually carried out show that achieved Defined Cost and timing are different from the forecasts included in the accepted quotation or in the Project Manager's assessment, the assessment is not changed. This gives the implementation of the compensation event *finality*.

There is a mistaken belief that Clause 60.1(8) "the Project Manager or the Supervisor changes a decision which he has previously communicated to the Contractor", allows the Project Manager to "undo" his assessment of a compensation event, i.e. to change his mind at a later date. This is incorrect – the word "decision" like the word "implementing" has a specific meaning under NEC4 contracts.

The Project Manager assessing and implementing a compensation event is not the same as giving a decision. It must be stressed that the Project Manager does not have the authority to change his assessment once it has been implemented. The Project Manager's assessment is final and the only recourse would be adjudication between the Parties, i.e. the Client and the Contractor.

Chapter 7

Title

Title to Plant and Materials, objects and materials within the Site

Question 7.1 During excavations for the foundations to a new building as part of a project to build a new school using the NEC4 Engineering and Construction Contract Option A (priced contract with activity schedule), the Contractor has discovered some archaeological remains. How should this discovery be dealt with under the contract?

If the Contractor discovers any object of value or of other interest, first it is important to recognise that he has no title to it (Clause 73.1).

In this particular case, dealing with archaeological remains, especially where human remains are also found, is a very specialist and prescribed process as there will normally be specific regulations, and also, dependant on where the Site is, statutory requirements, regarding dealing with the discovery, allowing other third parties to examine what has been found before carrying out any further work, whether they should then be preserved in situ or removed from Site, and who has title to what is discovered.

This can also have a severe impact on the programme and the cost of the project so it is vital that all possible Site and other investigations are carried out at the time of preparing the tender documents so that the Employer, the Contractor, the Project Manager and any other parties are as aware as they can be of the likelihood of finding such remains and how to deal with them.

Obviously, in the case of a school, which may have to be completed in time for the start of a new academic term, this could be a serious issue, though on a positive note, the find could provide a great opportunity for the schoolchildren to be made aware of the importance of protecting our archaeological heritage!

Going back to the contract, under Clause 73.1 of the NEC4 Engineering and Construction Contract, the Contractor must notify the Project Manager upon finding anything as described in the question, who then instructs the Contractor how to deal with it. It is vital that the Contractor only takes instructions from the Project Manager in this respect, although in the case of archaeological remains, other Parties such as local authorities, museums and educational establishments may have an interest in the find and how it is dealt with.

Initially, an early warning notice should be issued by the Contractor (or the Project Manager) allowing the parties to share their thoughts and opinions on the possible impact upon Price and Completion Date and probably an early warning meeting should be held to consider proposals, seek solutions and decide actions.

This is a compensation event under Clause 60.1(7), so it is important that the Contractor not only notifies the Project Manager of a compensation event, if the Project Manager has not already notified, but also keeps detailed records of any stoppages, disruption and any changes to working methods as a result of the find, as he will need to substantiate the effect on Defined Cost, and also on any delay to the Completion Date.

Chapter 8

Indemnity, insurance and liability
Insurance requirements, claims, etc.

Question 8.1 What are the insurance requirements within the NEC4 contracts?

In order to consider this question, let us consider three of the main NEC4 contracts:

- NEC4 Engineering and Construction Contract
- NEC4 Professional Service Contract
- NEC4 Term Service Contract

NEC4 Engineering and Construction Contract

Clause 80.1 lists the Client's liabilities; however, these are the Client's liabilities only in terms of loss, wear or damage, and are not an exhaustive list of all the liabilities the Clients bears under the contract. There are many other liabilities such as actions or inactions of the Project Manager or Supervisor, unforeseen physical conditions and weather, which are covered elsewhere as compensation events.

There are nine main categories of Client's liability within this clause:

1. Claims and proceedings from Others and costs payable to Others relating to the Client's use or occupation of the Site, negligence, breach of statutory duty or interference with any legal right by the Client or anyone employed or contracted to it, except the Contractor
2. A fault of the Client or any person employed by or contracted to it, except the Contractor
3. A fault in the design contained in the Scope provided by the Client or an instruction from the Project Manager changing the Scope

    In respect of design, the Client should cover the risk itself through professional indemnity insurance, or if it uses external consultants for the design, it should ensure that they hold such insurance. This insurance should be held for the full period of liability, which will normally be several years post completion of the works.

4   Loss or damage to Plant or Materials supplied by the Client to the Contractor or by Others on the Client's behalf until the Contractor has received and accepted them
5   Loss or damage to the works, Plant and Materials, caused by matters outside the control of the Parties
6   Loss or damage to parts of the works taken over by the Client, except where due to a Defect that existed at takeover, an event occurring before takeover which was not a Client's liability, or due to the activities of the Contractor on the Site after takeover
7   Loss or damage to the works and any Equipment, Plant and Materials retained on Site by the Client after termination, except loss or damage due to the activities of the Contractor on the Site after termination
8   Loss or damage to property owned or occupied by the Client other than the works, unless arising from or in connection with the Contractor Providing the Works
9   Additional Client's liabilities stated in the Contract Data.

Clause 81.1 lists the Contractor's liabilities:

1   Claims and proceedings from Others, and compensation and costs payable to Others which arise from the Contractor Providing the Works.
2   Loss of or damage to the works, Plant and Materials and Equipment.
3   Loss of or damage to property owned or occupied by the Client other than the works which arise from the Contractor Providing the Works.
4   Death or bodily injury to employees of the Contractor.

Recovery of Costs

- any cost which the Client pays as a result of an event for which the Contractor is liable is paid by the Contractor
- any cost which the Contractor pays as a result of an event for which the Client is liable is paid by the Client.

The right to recovery of such costs is reduced if the recovering Party is partly liable for the costs.

The Insurance Table

The Insurance Table in the contract itemises the insurances that the Contractor has to effect, together with the minimum amount of cover.

The default position is that the Contractor provides, maintains and pays for the insurances, unless the Client states otherwise in Contract Data Part 1.

Whether the Contractor or the Client takes out the insurances, they are always effected as a joint names policy, except the fourth insurance, which is

Employer's Liability Insurance. The insurances are to be effective from the starting date until the Defects Certificate or a termination certificate has been issued.

The four insurances listed within the Insurance Table are:

1. Loss or damage to the works, Plant and Materials

    Insurance of the Works will normally be covered by the Contractor's All Risks (CAR) policy. The *minimum* amount of cover for all insurances is stated in the Insurance Table, however the Contractor is liable for *whatever* the amount of any claim, therefore it must consider the minimum value in the Insurance Table purely as a guide.

2. Loss or damage to Equipment

    Again, the Contractor's All Risks (CAR) policy should cover this. The reference to "replacement cost" means the cost of replacement with Equipment of similar age and condition rather than "new for old".

3. Loss or damage to property (except the works, Plant or Materials) or bodily injury or death of an employee of the Contractor arising in connection with the Contractor providing the Works.

    This requires the Contractor to indemnify the Client against any loss, expense, claim, etc. in respect of any personal injury or death caused by the carrying out of the work, other than their own employees. This includes the liability toward members of the public who may be affected by the construction work, although they have no part in it. In the case where a party makes a claim directly against the Client due to a death or injury the Contractor should either take on that claim, or alternatively the Client can sue the Contractor to recover any monies.

    The Insurance Table states the minimum amount of cover or minimum limit of indemnity. The Contractor could be liable for whatever the amount of any Claim, therefore it must consider the minimum value in the Insurance Table purely as a guide.

4. Death of, or bodily injury to, employees of the Contractor

    As stated above, this is the only insurance within the Insurance Table that is not effected in the joint names of the Contractor and the Client. In many countries this insurance is a legal requirement for a company that employs people.

    This covers the Contractor's liabilities as an employer of people to insure against injury or death caused to those employees whilst carrying out their work, which is a legal obligation in most countries.

    Again, the Insurance Table states the minimum amount of cover or minimum limit of indemnity, the Contractor having to insure for whatever the amount of any claim.

Professional Indemnity (PI) Insurance

Contract Data Part 1 provides for the Client to insert a requirement for the Contractor to provide additional insurance.

Where the Contractor is designing parts of the works, it would be advisable for the Client to include a requirement for Professional Indemnity (PI) Insurance (See Question 8.2)

Proof of Insurance

The Contractor is required to submit to the Project Manager for acceptance certificates of insurance as required by the contract signed by the insurer or the insurance broker:

1   Before the starting date
2   On each renewal of the insurance policy.

If the Contractor does not submit a required certificate, the Client may insure a risk, the cost of the premium being paid by the Contractor.

If the Client is to provide insurance, then the Project Manager submits certificates to the Contractor for acceptance before the starting date and as and when instructed by the Contractor.

If the Client does not submit a required certificate, the Contractor may insure a risk, the cost of the premium being paid by the Client.

Professional Service Contract

The Professional Service Contract again states the Client's liabilities, this time there are only three:

1   Claims and proceedings from Others and compensation and costs payable to Others which are due to the unavoidable result of the service or negligence, breach of statutory duty or interference with any legal right by the Client or by anyone employed or contracted to it, except the Consultant.
2   A fault of the Client or any person employed by or contracted to it, except the Consultant.
3   Additional Client's liabilities stated in the Contract Data.

The Insurance Table covers:

1   Professional Indemnity Insurance
2   Public Liability Insurance
3   Employer's Liability Insurance.

Again, either Party is required to submit to the other certificates of insurance as required by the contract signed by the insurer or the insurance broker.

Term Service Contract

There are eight main categories of Client's liability within this clause:

1. Claims and proceedings from Others and costs payable to Others relating to the Client's use or occupation of the Site, negligence, breach of statutory duty or interference with any legal right by the Client or anyone employed or contracted to it, except the Contractor.
2. A fault of the Client or any person employed by or contracted to it, except the Contractor.
3. Loss or damage to equipment and Plant and Materials supplied to the Contractor by the Client, or by Others on the Client's behalf until the Contractor has received and accepted them.
4. Loss of or damage to Plant or Materials due to war, civil war, strikes, riots, radioactive contamination, etc.
5. Loss or damage to any Equipment, Plant and Materials retained by the Client after a termination, except loss or damage due to the activities of the Contractor in the Affected Property after the termination.
6. Loss or damage to Affected Property and any other property owned or occupied by the Client, unless the loss or damage arises from or in connection with the Contractor Providing the Service.
7. Loss or damage to any Plant and Materials after they have been included in the Affected Property.
8. Additional Client's liabilities stated in the Contract Data.

The Insurance Table covers:

1. Loss or damage to Plant and Materials and Equipment
2. Public Liability Insurance
3. Employer's Liability Insurance.

Again, either Party is required to submit to the other certificates of insurance as required by the contract signed by the insurer or the insurance broker.

Question 8.2 Can we include a requirement in an NEC4 contract for a Contractor or a Consultant to have Professional Indemnity (PI) insurance?

Let us first consider the subject of Professional Indemnity (PI) insurance.

If a Contractor or Consultant providing a service to a Client makes a mistake, is found to be negligent, or gives inaccurate advice, then he will be liable to the Client in event that the Client incurs a loss as a result. This loss can be very significant where the design has to be corrected, parts of the structure have to be taken down and reinstated, a facility has to be closed down whilst the remedial measures take place, and there are also legal costs.

Professional Indemnity claims can arise where there is negligence, misrepresentation or inaccurate advice which does not give rise to bodily injury, property damage or personal injury, but does give rise to some financial loss. Additional coverage for breach of warranty, intellectual property, personal injury, security and cost of contract can be added. In that event, although the Client claims against the Contractor or Consultant rather than from the insurers, PI Insurance protects the Contractor or Consultant against claims for loss or damage made by a client or third party.

If we first consider the NEC4 Professional Service Contract, within Clause 83.3, the Insurance Table includes the *"liability of the Consultant for claims made against him arising out of his failure to use the skill and care normally used by professionals providing services similar to the service"*, i.e. the services defined under the specific contract – this is Professional Indemnity (PI) insurance.

The minimum amount of cover is then stated within Contract Data Part 1 together with the period following Completion of the whole of the services which the insurance must be maintained for, unless there is an earlier termination.

In respect of the period for which the insurance is maintained, it is very important to mention that Professional Indemnity insurance policies are based on a "claims made" basis, meaning that the policy only covers claims made during the policy period when the policy is "live", so claims that may relate to events occurring before the coverage was active may not be covered.

However, these policies may have a retroactive date that can operate to provide cover for claims made during the policy period but that relate to an incident after the retroactive date.

The Service Manager, on behalf of the Client, should ensure that the Consultant has taken out and maintained the required insurances for the full period of its liability. Claims that may relate to incidents occurring before the policy was active may not be covered, although a policy may sometimes have a retroactive date.

By default, the NEC4 construction contracts, where a party is the Contractor, i.e. the Engineering and Construction Contracts and the Term Service Contracts, do not specifically require the Contractor to have Professional Indemnity (PI) insurance, but Contract Data Part 1 provides for the Client to insert a requirement for the Contractor to provide any additional insurances, which can include Professional Indemnity (PI) insurance.

Particularly where the Contractor is designing parts of the works, it would certainly be advisable for the Client to include a requirement for the Contractor to provide Professional Indemnity (PI) Insurance, as although any Consultant employed by the Contractor may have relevant Professional Indemnity insurance, the liability to the Client will lie with the Contractor, not with the Consultant he employs.

Chapter 9

Termination provisions

Reasons, procedures and amounts due

Question 9.1   We are Contractors on an NEC4 Engineering and Construction Contract Option A (priced contract with activity schedule), and we wish to terminate our employment under the contract due to non-payment by the Client. Can we do this, and if so how?

The first thing to say is that termination should never be a decision taken lightly by *any* Party under *any* contract – it really should be an action of last resort when the Parties are not able to amicably resolve issues between themselves.

Termination is a word most commonly used in the context of construction contracts to refer to the ending of the Contractor's employment. The Parties have a common law right to bring the contract to an end in certain circumstances, but most standard forms give the Parties additional and express rights to terminate upon the happening of specified events.

Some contracts refer to termination of the contract, whilst others refer to the termination of the Contractor's employment under the contract. In practice, it makes little difference, and most contracts make express provision for what is to happen after termination.

Also, termination is an entitlement not an obligation, though clearly, if one of the Parties has become insolvent, then there is no choice but to terminate the contract.

It is also important to note that neither Party should attempt to terminate employment under the contract unless they are sure that the provision is available within the contract and if it is, then they must ensure that they strictly comply with the wording of the contract. If termination is held to be wrongful, it is usually a repudiation of the contract.

It should be noted that most contracts require a notice to be given by one Party to the other unless the breach is due to insolvency; the offending Party then has a period of time in which to remedy the breach, failing which, termination can take place.

Termination, including reasons, termination procedures and payment are covered within the Engineering and Construction Contract by Clauses 90.1 to 93.2 with further references within the Main Option clauses.

Within the Engineering and Construction Contract, there are 22 reasons listed for either Party to terminate:

- Reasons R1 to R10 provide for either Party to terminate due to the insolvency of the other. This includes bankruptcy, appointment of receivers, winding-up orders and administration orders, dependent on whether the Party is an individual, a company or a partnership.
- Reasons R11 to R13 provide for the Client to terminate if the Project Manager has notified the Contractor that he has defaulted and the Contractor has not remedied the default within four weeks of the notification in respect of the Contractor substantially failing to comply with his obligations, not providing a bond or guarantee, or appointing a Subcontractor for substantial work before the Project Manager has accepted the Subcontractor.

  It is quite subjective as to what the words "substantially failed to comply" and "substantial work" means within these provisions. One would assume that it is not intended to relate to minor breaches, but having said that, the Contractor could "substantially fail to comply" with a minor obligation, or alternatively "fail to comply" with a major obligation.
- Reasons R14 to R15 provide for the Client to terminate if the Project Manager has notified that the Contractor has defaulted and the Contractor has not stopped defaulting within four weeks of the notification in respect of substantially hindering the Client or Others, or has substantially broken a health or safety regulation and not stopped defaulting within the four weeks of the notification.

  Again, the use of the word "substantially" is quite subjective. How would one substantially hinder someone, or substantially break a health or safety regulation? Surely, a health or safety regulation is either "broken" or "not broken"?
- Reason 16 provides for the Contractor to terminate if the Client has not paid an amount certified by the Project Manager within 13 weeks of the date the certificate should have been paid.
- Reason 17 provides for either Party to terminate if the Parties have been released under the law from further performance of the whole contract. Note the reference to "the whole contract", not just a part.
- Reasons 18 to 20 provide for the Project Manager having instructed the Contractor to stop or not restart any substantial work or all work and an instruction allowing the work to restart or start has not been given within 13 weeks. Either Party may terminate if the instruction was due to a default by the other, or if the instruction was due to any other reason.
- Reason R21 provides for the Client to terminate if an event occurs that stops the Contractor completing the works, or stops the Contractor completing the works by the date shown on the Accepted Programme and is forecast to delay Completion by more than 13 weeks, and which neither Party could prevent and an experienced Contractor would have judged at the Contract

Date to have such a small chance of occurring that it would have been unreasonable for him to have allowed for it.
- Reason R22 is a new reason under the NEC4 contracts and provides for the Client to terminate if the Contractor does a Corrupt Act, unless it was done by a Subcontractor or supplier, and the Contractor was not and should not have been aware of the Corrupt Act, or it informed the Project Manager of the Corrupt Act and took action to stop it as soon as it became aware of it.

Note that the Contractor can only terminate for one of the above reasons, though the Client can, under Secondary Option X11, terminate for a reason not stated in the Termination Table.

Whilst this may seem inequitable, if the Client terminates for a reason not stated in the Termination Table, in addition to paying the Contractor for all work carried out up to the termination, he will have to pay other costs incurred in the expectation of completing the whole of the works; for example, orders placed and other commitments made, the forecast Defined Cost of removing Equipment and, dependent on the Main Option used, the fee percentage applied to the difference between the original total of the Prices and the Price for Work Done to Date, essentially a loss of profit/overheads provision.

Clearly, the earlier the Client exercises this right of termination, the greater this amount. With all the termination reasons, the Party wishing to terminate notifies the Project Manager and the other Party, giving details of his reasons for terminating. The Project Manager then issues a termination certificate to both Parties if he is satisfied that the reason for termination is valid under the contract. It is perhaps curious that the Project Manager, who acts for the Client, makes the decision as to whether the termination is valid?

So, if the Contractor on an NEC4 Engineering and Construction Contract Option A wishes to terminate their obligation to Provide the Works under the contract due to non-payment by the Client, the first action is to check whether it is one of the reasons under the contract.

In this case, it is Reason 16 (R16), "*the Contractor may terminate if the Client has not paid an amount due under the contract within thirteen weeks of the date that it should have been paid*", so clearly the thirteen weeks is a critical factor in deciding whether the Contractor can terminate under the Contract.

So let us consider exactly how termination can be effected in the case of this Contractor.

First, the Contractor must notify the Project Manager and the Client that he wishes to terminate and give details of his reasons for terminating. If the reason complies with the Contract, then the Project Manager issues a termination certificate to the Contractor and to the Client.

After the termination certificate has been issued, the Contractor does no further work necessary to Provide the Works.

The procedures of termination are then implemented immediately upon issue of the termination certificate.

Procedures on termination (P1 and P4)

The Contractor leaves the Working Areas and removes his Equipment.

The Client may complete the works, either himself or using another Contractor, and may use any Plant and Materials to which he has title.

Payment on termination (A1, A2 and A4)

The Contractor is entitled to be paid:

- an amount due as for normal payments, in this case the Contractor has been employed under Option A of the Engineering and Contraction Contract
- the Defined Cost for Plant and Materials within the Working Areas, or to which the Client has title and of which the Contractor has to accept delivery
- other Defined Cost reasonably occurred by the Contractor in completing the works
- any amounts retained by the Client, and
- a deduction of any un-repaid balance of an advance payment

also,

- the forecast Defined Cost of removing the Equipment

also,

- for Options A, B, C and D, the fee percentage applied to any excess of the total of the Prices at the Contract Date (referred to in other contracts as the "Contract Sum") over the Price for Work Done to Date (referred to in other contracts as the "Gross Valuation").

So, if the fee percentage was 8% and the Price for Work Done to Date at termination is £2,500,000 and total of the Prices at the Contract Sum was £4,200,000, the amount payable by the Client to the Contractor is £136,000.

$$(£4,200,000 - £2,500,000) \times 8\% = £136,000$$

Interest on late payment

Note that with late or non-payments from the Client to the Contractor, under Clause 51.2 of the Engineering and Construction Contract, the Contractor is entitled to be paid interest which is assessed from the date by which the late payment should have been made until the date when the payment is made. This interest is calculated on a daily basis at the interest rate Contract Data Part 1 and is compounded annually.

Option Y(UK)2: The Housing Grants, Construction and Regeneration Act 1996

Note that under Option Y(UK)2 (if selected), which relates to the Housing Grants, Construction and Regeneration Act 1996 and the Local Democracy, Economic Development and Construction Act 2009, if payment is late, the Contractor may exercise his right under the Act to suspend performance due to late or non-payment.

It is also a compensation event, so the Contractor can recover any financial effect and/or delay to Completion as a result of exercising his right to suspend all or part of the works.

Chapter 10

Dealing with disputes
Adjudication and tribunal

Question 10.1 What is the difference between Option W1, W2 and W3 within the NEC4 Engineering and Construction Contract?

Resolving and avoiding disputes is covered within the Engineering and Construction Contract by Options W1, W2 and W3.

- Option W1 is used when Adjudication is the method of dispute resolution and the UK Housing Grants, Construction and Regeneration Act 1996 does not apply.
- Option W2 is used when Adjudication is the method of dispute resolution and the UK Housing Grants, Construction and Regeneration Act 1996 applies.
- Option W3 is used when a Dispute Avoidance Board is the method of dispute resolution and the UK Housing Grants, Construction and Regeneration Act 1996 does not apply.

Note: The Local Democracy, Economic Development and Construction Act 2009 which came into force in England and Wales on 1 October 2011 and in Scotland on 1 November 2011 has amended the Housing Grants, Construction and Regeneration Act 1996, the new Act applying to most UK contracts after that date.

If the Engineering and Construction Contract is used within the UK, the Client must, by reference to the Act, determine whether the contract is a "construction contract", and select the appropriate dispute resolution Option. If the Engineering and Construction Contract is used outside the UK, Option W1 or W3 will apply.

Clearly, there is nothing to prevent the Parties jointly attempting to resolve a dispute as they see fit, by a consensual and non-binding route such as negotiation, mediation, conciliation or mini trial, prior to referring it to adjudication.

Option W1

As stated above, Option W1 is used when the UK Housing Grants, Construction and Regeneration Act 1996 does not apply.

Any dispute arising under the contract is, in accordance with the Dispute Reference Table, first referred by a Party issuing a notice to the Senior Representatives and copied to the other Party and the Project Manager, stating the nature of the dispute it wishes them to resolve.

Each Party then submits their statement of case (no more than ten sides of A4 paper) with supporting evidence.

The Senior Representatives then try to resolve the dispute within three weeks, using any procedure they wish to and attending as many meetings as required to resolve the dispute. At the end of the three weeks, a list is issued showing the issues agreed, and the issues not agreed. The Project Manager and the Contractor put the agreed issues into effect.

According to the Dispute Reference Table, if the Dispute is about:

- an action or inaction of the Project Manager or the Supervisor, either Party may refer it to the Senior Representatives, not more than four weeks after the Party became aware of the action or inaction
- a programme, compensation event, or quotation for a compensation event which is treated as having been accepted, the Client may refer it to the Senior Representatives, not more than four weeks after it was treated as accepted
- an assessment of Defined Cost which is treated as correct, either Party may refer it to the Senior Representatives, not more than four weeks after the assessment was treated as correct
- any other matter, either Party may refer it to the Senior Representatives, when the dispute arises.

Note that if the referring Party is the Client and the dispute is regarding a compensation event which has been treated as having been accepted (Clause 62.6) then the Client must notify the dispute to the Contractor and then refer it to the Senior Representatives not more than four weeks after it was treated as accepted.

This seems an odd reason for referring a dispute to the Senior Representatives, as Clause 62.6 states that if the Project Manager does not reply to the Contractor's notification, it is treated as acceptance of the quotation; how could the Senior Representatives (or an Adjudicator) whose role is to enforce the contract find other than in favour of the Contractor?

Surely the Client's dissatisfaction, and the dispute itself, is with its Project Manager, not with the Contractor?

Whilst these timescales are fixed within the Dispute Reference Table, they can be extended by the Project Manager if the Contractor and the Project Manager agree before either the notice or the referral is due. Note that if the matter in dispute is not notified and referred within these timescales, neither Party can refer the dispute to the Adjudicator or the tribunal.

The Adjudicator is appointed under the NEC4 Dispute Resolution Service Contract. The Adjudicator acts impartially, and if it resigns or is unable to act, the Parties jointly appoint a new Adjudicator. The referring Party obtains a copy and completes the Dispute Resolution Service Contract. If the Parties have not appointed an Adjudicator, either Party may ask the nominating body to choose an Adjudicator.

No procedures have been specified for appointing a suitable person, and in practice a number of different methods have been used. Whatever method is used, it is important that both Parties have full confidence in its impartiality, and for that reason it is preferable that a joint appointment is made.

The Adjudicator should be a person with experience in the type of work included in the contract between the Parties and who occupies or has occupied a senior position dealing with disputes. He should be able to listen to and understand the viewpoint of both Parties.

Often the Parties delay selecting an Adjudicator until a dispute has arisen, although this frequently results in a disagreement over who should be the Adjudicator.

As noted, the selection of the Adjudicator is important, and it should be recognised that a failure to agree an Adjudicator means that a third party will make the selection without necessarily consulting the Parties.

The adjudication process commences by the referring Party issuing a notice of adjudication to the Project Manager, and to the other Party within two weeks of the production of the list of agreed and not agreed issues.

The notice of adjudication gives a brief description of the dispute, details of where and when the dispute has arisen, and the nature of the redress that is sought.

The Party referring the dispute to the Adjudicator must include within its referral information that it wishes to be considered by the Adjudicator, any more information from either Party to be provided within four weeks of the referral.

A dispute under a subcontract, which is also a dispute under the TSC, may be referred to the Adjudicator at the same time, and the Adjudicator can decide the two disputes together.

The Adjudicator may review and revise any action or inaction of the Project Manager, take the initiative in ascertaining the facts, and the law relating to the dispute, instruct a Party to provide further information and instruct a Party to take any other action that it considers necessary to reach its decision.

All communications between a Party and the Adjudicator must be communicated to the other Party at the same time.

The Adjudicator decides the dispute and notifies the Parties and the Project Manager within four weeks of the end of the period for receiving information.

This four-week period may be extended by joint agreement between the Parties.

Until this decision has been communicated, the Parties proceed as if the matter in dispute was not disputed.

The Adjudicator's decision is temporarily binding on the Parties unless and until revised by a tribunal and is enforceable as a contractual obligation on the Parties.

The Adjudicator's decision is final and binding if neither Party has notified the Adjudicator that they are dissatisfied with an Adjudicator's decision within the time stated in the contract, and intends to refer the matter to the tribunal.

The Adjudicator may, within two weeks of giving its decision to the Parties, correct a clerical mistake or ambiguity.

Review by the tribunal

A dispute cannot be referred to the tribunal unless it has first been referred to the Adjudicator.

The tribunal will be named by the Client within Contract Data Part 1.

Whilst no alternatives are stated, it will normally be legal proceedings or arbitration.

If arbitration is chosen, the Client must also state in Contract Data Part 1, the procedure, the place where the arbitration is to be held, and the person or organisation who will choose an Arbitrator if the Parties cannot agree a choice, or if the named procedure does not state who selects the Arbitrator.

A Party can, following the adjudication, notify the other Party within four weeks of the Adjudicator's decision that it is dissatisfied. This is a time-barred right, as the dissatisfied Party cannot refer the dispute to the tribunal unless it is notified within four weeks of the Adjudicator's decision; failure to do so will make the Adjudicator's decision final and binding.

Also, if the Adjudicator has not notified its decision within the time provided by the contract, a Party may within four weeks of when the Adjudicator should have given its decisions notify the other Party that it intends to refer the dispute to the tribunal. The tribunal settles the dispute and has the power to reconsider any decision of the Adjudicator and review and revise any action or inaction of the Project Manager.

It is important to note that the tribunal is not a direct appeal against the Adjudicator's decision; the Parties have the opportunity to present further information or evidence that was not originally presented to the Adjudicator, and also the Adjudicator cannot be called as a witness.

Option W2

As stated above, Option W2 is used when the UK Housing Grants, Construction and Regeneration Act 1996 applies.

The inclusion of adjudication within the then New Engineering Contract, now the NEC4, predates UK legislation giving Parties to a contract the statutory right to refer a dispute to adjudication.

The Housing Grants, Construction and Regeneration Act 1996

The Local Democracy, Economic Development and Construction Act 2009, which came into force in England and Wales on 1 October 2011 and in Scotland on 1 November 2011, amended the Housing Grants, Construction and Regeneration Act 1996, the new Act applying to most UK contracts after that date.

There still remain certain categories of construction contract to which neither Act applies.

The differences between the Housing Grants, Construction and Regeneration Act 1996 and the Local Democracy, Economic Development and Construction Act 2009 have been explored in a previous chapter but in respect of adjudication they include the following:

- The previous Act only applied to contracts that were in writing; the new Act also applies to oral contracts.
- Terms in contracts such as "the fees and expenses of the Adjudicator as well as the reasonable expenses of the other party shall be the responsibility of the party making the reference to the Adjudicator" will be prohibited.

  Much has been written about the effectiveness of such clauses, and whether they comply with the previous Act, so the new Act should provide clarity for the future.
- The Adjudicator is permitted to correct its decision so as to remove a clerical or typographical error arising by accident or omission. Previously it could not make this correction.
- Section 108 of the Act provides Parties to construction contracts with a right to refer disputes arising under the contract to adjudication. It sets out certain minimum procedural requirements that enable either Party to a dispute to refer the matter to an independent Party who is then required to make a decision within 28 days of the matter being referred.
- If a construction contract does not comply with the requirements of the Act, or if the contract does not include an adjudication procedure, a statutory default scheme, called the Scheme for Construction Contracts (referred to as the "Scheme"), will apply.

The Act provides that a dispute can be referred to adjudication "at any time" provided the Parties have a contract.

"At any time" can also refer to a dispute after the contract is completed.

What are the requirements within construction contracts?

Section 108 of the Act requires all "construction contracts", as defined by the Act, to include minimum procedural requirements that enable the Parties to a contract to give notice at any time of an intention to refer a dispute to an Adjudicator.

The contract must provide a timetable so that the Adjudicator can be appointed, and the dispute referred, within seven days of the notice. The Adjudicator is required to reach a decision within 28 days of the referral, plus any agreed extension, and must act impartially. In reaching a decision, an Adjudicator has wide powers to take the initiative to ascertain the facts and law related to the dispute.

The Housing Grants, Construction and Regeneration Act 1996 defines adjudication as:

> *a summary non judicial dispute resolution procedure that leads to a decision by an independent person that is, unless otherwise agreed, binding upon the parties for the duration of the contract, but which may subsequently be reviewed by means of arbitration, legal proceedings or by agreement.*

In that sense, adjudication does not necessarily achieve final settlement of a dispute because either of the Parties has the right to have the same dispute heard afresh in court, or where the contract specifies arbitration.

However, experience since the Housing Grants, Construction and Regeneration Act 1996 came into force shows that the majority of adjudication decisions are accepted by the Parties as the final result.

For Option W2, there is no Dispute Reference Table as the Parties have a right to refer a dispute to each other and to the Adjudicator at any time.

Any dispute arising under the contract is first referred by a Party issuing a notice to the Senior Representatives and copied to the other Party and the Project Manager stating the nature of the dispute it wishes them to resolve. Each Party then submits their statement of case (no more than ten sides of A4 paper) with supporting evidence.

The Senior Representatives then try to resolve the dispute within three weeks, using any procedure they wish to. At the end of the three weeks, a list is issued showing the issues agreed, and the issues not agreed. The Project Manager and the Contractor put the agreed issues into effect.

The requirements of the Housing Grants, Construction and Regeneration Act 1996 for adjudication to be possible at any time means that the involvement of the Senior Representatives may be bypassed by Parties and the dispute referred straight to the adjudicator.

Again, as with Option W1, the Adjudicator is appointed under the NEC4 Dispute Resolution Service Contract. The Adjudicator acts impartially, and if it resigns or is unable to act, the Parties jointly appoint a new Adjudicator. The referring Party obtains a copy and completes the Dispute Resolution Service Contract. If the Parties have not appointed an Adjudicator, either Party may ask the nominating body to choose an Adjudicator.

A Party may first give a notice of adjudication to the other Party with a brief description of the dispute, details of where and when the dispute has arisen, and the nature of the redress that is sought.

The Adjudicator may be named in the contract in which case the Party sends a copy of the notice to the Adjudicator. The Adjudicator must confirm within three days of receipt of the notice that it is able to decide the dispute, or if it is unable to decide the dispute.

Within seven days of the issue of the notice of adjudication, the Party:

- refers the dispute to the Adjudicator
- provides the Adjudicator with the information on which it relies together with supporting information
- provides a copy of the information and supporting documents to the other Party.

Again, a dispute under a subcontract, which is also a dispute under the TSC, may be referred to the Adjudicator at the same time and the Adjudicator can decide the two disputes together.

The Adjudicator may review and revise any action or inaction of the Project Manager, take the initiative in ascertaining the facts, and the law relating to the dispute, instruct a Party to provide further information and instruct a Party to take any other action that it considers necessary to reach its decision.

All communications between a Party and the Adjudicator must be communicated to the other Party at the same time.

The Adjudicator decides the dispute and notifies the Parties and the Project Manager within 28 days of the dispute being referred.

This 28-day period may be extended by 14 days with the consent of the referring Party or by any other period by joint agreement between the Parties.

Until this decision has been communicated, the Parties proceed as if the matter in dispute was not disputed.

The Adjudicator's decision is temporarily binding on the Parties unless and until revised by a tribunal and is enforceable as a contractual obligation on the Parties.

The Adjudicator's decision is final and binding if neither Party has notified the Adjudicator that they are dissatisfied with an Adjudicator's decision within the time stated in the contract, and intends to refer the matter to the tribunal.

The Adjudicator may, within five days of giving its decision to the Parties, correct any clerical mistake or ambiguity.

Review by the tribunal

A dispute cannot be referred to the tribunal unless it has first been referred to the Adjudicator.

Again, the tribunal will be named by the Client within Contract Data Part 1, normally legal proceedings or arbitration. If arbitration is chosen, the Client must also state in Contract Data Part 1, the procedure, the place where the arbitration is to be held, and the person or organisation who will choose an Arbitrator if the Parties cannot agree a choice, or if the named procedure does not state who selects the Arbitrator.

A Party can, following the adjudication, notify the other Party within four weeks of the Adjudicator's decision that it is dissatisfied. This is a time-barred right, as the dissatisfied Party cannot refer the dispute to the tribunal unless it is notified within four weeks of the Adjudicator's decision; failure to do so will make the Adjudicator's decision final and binding.

The tribunal settles the dispute and has the power to reconsider any decision of the Adjudicator and review and revise any action or inaction of the Project Manager.

It is important to note that the tribunal is not a direct appeal against the Adjudicator's decision; the Parties have the opportunity to present further information or evidence that was not originally presented to the Adjudicator, and also the Adjudicator cannot be called as a witness.

Option W3

Published construction contracts in the past have usually included some form of dispute resolution provisions including legal proceedings, arbitration, dispute adjudication boards, adjudication (by contract and/or by statute), expert determination and mediation. In the absence of such provisions, or a mutual agreement between the Parties to adopt their own methods, the default would be for the Parties to refer to legal proceedings.

In recent years, contracting Parties, and in turn contract drafters, have also looked towards dispute *avoidance* to try to prevent issues escalating into dispute in the first place. This can be in the form of more balanced contract conditions, partnering agreements, target contracts, but also by including specific dispute avoidance provisions.

Option W3 within the NEC3 Engineering and Construction Contract is such a provision, and is a new addition to the NEC4 contracts, which is only included in some of the contracts. Note that W3 is an Option for *avoiding* a dispute, *not resolving* it.

Option W3 is used when a Dispute Avoidance Board is the method of dispute resolution and the United Kingdom Housing Grants Construction and Regeneration Act 1996 does not apply.

The Dispute Avoidance Board consists of one or three members (odd number) as identified within the Contract Data Part 1. If the Contract Data states three members, the Parties jointly choose the third member.

The Dispute Avoidance Board is appointed using the NEC4 Dispute Resolution Service Contract (see above).

If a member of the Dispute Avoidance Board is not named within the Contract Data or is unable to act, then the Parties jointly choose a replacement. If the Parties do not do so, the Dispute Avoidance Board may do so themselves within seven days of the request to do so.

The Contract Data states the frequency of visits to Site by the Dispute Avoidance Board from the starting date to the defects date, but may make further visits if requested by the Parties.

The members of the Dispute Avoidance Board, their employees or agents, are not liable for any action or failure to take action to resolve a potential dispute, unless the failure was in bad faith.

A potential dispute under the contract may be referred to the Dispute Avoidance Board, two to four weeks after notification to the other Party and to the Project Manager.

The Parties are required to make all relevant information available to the Dispute Avoidance Board, who visits the Site (where applicable) and helps the Parties to settle the potential dispute without formal referral as a dispute.

Review by the tribunal

A dispute cannot be referred to the tribunal unless it has first been referred to the Dispute Avoidance Board as a potential dispute.

A Party can, following the Dispute Avoidance Board's recommendation, notify the other Party that it is dissatisfied. This is a time-barred right, as the dissatisfied Party cannot refer the dispute to the tribunal unless it is notified within four weeks of the Dispute Avoidance Board's recommendation.

The tribunal settles the dispute and has the power to reconsider any recommendation of the Dispute Avoidance Board and review and revise any action or inaction of the Project Manager.

It is important to note that the tribunal is not a direct appeal against the Dispute Avoidance Board's recommendation; the Parties have the opportunity to present further information or evidence that was not originally presented to the Dispute Avoidance Board, and also the Dispute Avoidance Board cannot be called as a witness.

Question 10.2 When are we required to select and name the Adjudicator on our NEC4 Engineering and Construction Contract?

Within Contract Data Part 1, there are three options for selecting and naming the Adjudicator:

- The name and address of the Adjudicator may be stated by the Client in Contract Data Part 1; the Adjudicator is then appointed before the starting date.

    This has the advantage that tendering Contractors are aware at the time of tender who will be the Adjudicator in the event that a dispute arises and, if necessary, can object to them within their tenders.

    Also, the Adjudicator is already in place should a dispute arise. The named Adjudicator may require some form of retainer fee for being named in the contract and being available in the event that a dispute arises.

    or
- The Parties can mutually agree to the name of the Adjudicator in the event that a dispute arises.

This has the advantage that there is no "Adjudicator in waiting" and therefore no fee payable. Many NEC practitioners say that pre-appointing the Adjudicator as in Option 1 is almost resigning oneself to the fact that there will at some point be a dispute.

However, once Parties are in dispute they then have to agree who is to be the Adjudicator and appoint him; however, at this point they may not wish to agree with each other about anything!

or

- The name of the Adjudicator nominating body may be stated by the Client in Contract Data Part 1.

This is probably the favoured option as an independent name can be put forward by the nominating body, who is normally well prepared to put forward the name of an Adjudicator and have him appointed within the timescales set by the contract and if necessary the appropriate legislation.

In all cases, the Adjudicator must be impartial, i.e. the Adjudicator should not show any bias towards either Party. In addition, all correspondence from the Adjudicator must be circulated to both Parties. Any request for an Adjudicator must be accompanied by a copy of the notice of adjudication, and the appointment of the Adjudicator should take place within seven days of the submission of the notice of adjudication to the other Party.

Adjudicator Nominating Bodies, as the name suggests, are organisations that fulfil the role of nominating Adjudicators. These bodies keep registers of Adjudicators with varying expertise and based at various geographical locations who can act for Parties should they be nominated.

Below is a list of recognised Adjudicator nominating bodies. It is not an exhaustive list, but clearly shows the wide range of organisations that can nominate Adjudicators.

- Association of Independent Construction Adjudicators (AICA)
- Association for Consultancy and Engineering (ACE)
- Building Disputes Tribunal of New Zealand
- Centre for Effective Dispute Resolution (CEDR)
- Chartered Institute of Arbitrators (CIArb)
- Chartered Institute of Arbitrators (Scotland) (CIArb-Scotland)
- Chartered Institute of Building (CIOB)
- Construction Conciliation Group (CCG)
- Construction Confederation (CC)
- Construction Industry Council (CIC)
- Construction Plant-Hire Association (CPA)
- Dispute Board Federation (DBF)
- Dispute Resolution Board Foundation (DRBF)
- Institution of Chemical Engineers (IChemE)
- Institution of Civil Engineers (ICE)

- ICE-SA
- Institution of Electrical Engineers (IEE)
- Institution of Mechanical Engineers (IMechE)
- Law Society of Scotland (LawSoc (Scot))
- Nationwide Academy of Dispute Resolution (NADR)
- RICS-Dispute Resolution Service (RICS-DRS)
- RICS Dispute Resolution Service Australia (RICS DRS (Oceania))
- Royal Incorporation of Architects in Scotland (RIAS)
- Royal Institute of British Architects (RIBA)
- Royal Institution of Chartered Surveyors (RICS)
- Royal Society of Ulster Architects (RSUA)
- Technology and Construction Court Bar Association (TECBAR)
- Technology and Construction Solicitors Association (TeCSA)
- UK Adjudicators.

Clearly, in naming the Adjudicator Nominating Body in the contract it is advisable to name an organisation with expertise in the project to be carried out. These bodies are very knowledgeable about appointment of Adjudicators and the relevant timescales and for a modest fee can nominate a suitably qualified Adjudicator to suit the Parties' requirements.

Question 10.3 What is the NEC4 Dispute Resolution Service Contract, and when should we use it?

The NEC4 Dispute Resolution Service Contract replaced the NEC3 Adjudicator's Contract.

The first edition of the NEC Adjudicator's Contract was published in 1994, and was written for the appointment of an Adjudicator for any contract under the NEC family of standard contracts. The second edition, published in 1998, contained some changes including the need to harmonise with the NEC standard contracts and further editions which had been issued since 1994. The third edition harmonised the contract with the NEC3 family of contracts using either Option W1 or W2.

The NEC4 Dispute Resolution Service Contract is made up of five parts:

1. General
2. Adjudication
3. Dispute Avoidance Board
4. Payment
5. Termination.

The Dispute Resolution Service Contract can also be used with contracts other than NEC4, though dependent on the contract and the applicable law, some amendment may be necessary.

In the UK, the Housing Grants, Construction and Regeneration Act 1996 has made adjudication mandatory as a means of resolving disputes in construction contracts that fall under the Act. Parties to a contract that does not provide for adjudication as required by the Act have a right to adjudication under the "Scheme for Construction Contracts". Schemes that are substantially similar have been published for England and Wales, Scotland and Northern Ireland.

The agreement is between the two disputing Parties and the Dispute Resolver, who may also be acting as the Adjudicator.

The contract contains five sections:

1   General

This covers the obligation upon the Dispute Resolver to act impartially, and to notify the Parties as soon as it becomes aware of any matter that could present a conflict of interest.

There is a definition of expenses, which includes printing costs, postage, travel, accommodation and the cost of any assistance with the adjudication. All communications must be in a form that can be read, copied and recorded. This is a requirement of all NEC contracts, but also of the adjudication process itself, with both Parties and the Dispute Resolver having to be copied into any communication between the Parties.

2   Adjudication

This Clause only applies if the Dispute Resolver is acting as an Adjudicator.

The Dispute Resolver cannot decide any dispute that is the same or substantially the same as its predecessor decided. It must make its decision and notify the Parties in accordance with the contract, and in reaching its decisions it can obtain assistance from others, but must advise the Parties before doing so, and also provide the Parties with a copy of anything produced by the assisting Party, so that the Parties can be invited to comment.

The Dispute Resolver's decision is to remain confidential between the Parties, and following its decision the Dispute Resolver retains documents provided to him for the period of retention, which is stated in the Contract Data.

An invoice is issued by the Dispute Resolver:

- each time a dispute is referred if an advance payment is stated in the Contract Date
- after a decision had been notified to the Parties;
- after a termination.

3   Dispute Avoidance Board

This Clause only applies if the Dispute Resolver is acting as a Dispute Avoidance Board member.

The Dispute Resolver carries out the duties of a Dispute Avoidance Board member in accordance with the contract between the Parties and collaborates with other members of the Dispute Avoidance Board.

4   Payment

The Dispute Resolver's hourly fee is stated in the Contract Data. There is provision for an advance payment to be made by the referring Party to the Dispute Resolver if stated in the Contract Data.

Following its decision, the Dispute Resolver submits an invoice to each Party for their share of the amount due, including expenses. The Parties pay the amount due within three weeks of receiving the Dispute Resolver's invoice.

Interest is paid at the rate stated in the Contract Data on any late payment. The Parties are jointly and severally liable to pay the amount due to the Dispute Resolver; therefore, if one Party fails to pay, the other must pay including any interest on the overdue amount, and recover the amount from the other Party.

5   Termination

The Parties may, by agreement, terminate the Dispute Resolver's agreement.

Also, the Dispute Resolver may terminate the agreement if there is a conflict of interest, it is unable to decide a dispute, an advance payment has not been made, or it has not been paid an amount due within five weeks of the date by which the payment should have been made. The Dispute Resolver's appointment terminates on the date stated in the Contract Data.

Question 10.4   In an NEC4 contract, what is the tribunal?

Whilst Contract Data Part 1 requires the Client to enter the form of the tribunal, it does not say what choices are available. There are various means of finally resolving disputes under a construction contract dependent on client choices and relevant legislation within the country using the NEC4 contracts, but the Client normally selects arbitration or legal proceedings.

If arbitration is chosen, within the Contract Data, the Client is required to state the arbitration procedure, the place where the arbitration is to be held, and who will choose an Arbitrator.

There are a number of differences between the two but essentially arbitration is hearing the dispute in private and legal proceedings is hearing the dispute in a public court.

It must be stressed that the role and authority of the tribunal is not to review and appeal against the previous decision of an Adjudicator, but to resolve the dispute independently of the adjudication; therefore, it is quite in order for Parties to introduce new evidence or materials to the tribunal.

Arbitration

Any dispute between two or more Parties can be resolved through the courts, known as legal proceedings.

However, legal proceedings have many disadvantages, not least the cost of a full court hearing, but in some cases a lengthy waiting time before the matter actually reaches the courts, so arbitration has become increasingly used as a simpler, more convenient method of final dispute resolution.

Arbitration is the hearing of a dispute by a third party, who is often not a lawyer, though that term has a very wide meaning, but is an expert in the field in which the dispute is based, and can give a decision based on opinion, which is then legally binding on both Parties.

Arbitration is not a new concept, in fact it has been in existence for almost as long as the law itself, the first official recognition in the UK being the Arbitration Act 1697, which largely governed disputes about the sale of livestock in cattle markets. With the continuing growth of freight transport by ship, and the advent of railways and other forms of transport, a large number of arbitration cases ensued, and as a result Parliament passed the Arbitration Act 1889.

Various amendments have been introduced over the following years with the current legislation being the Arbitration Act 1950, as amended by the Arbitration Acts of 1975 and 1979. The Arbitration Act 1996 is the current legislation related to arbitration in England and Wales.

One does not issue writs in arbitration as one would with legal proceedings; *both* Parties agree to enter into arbitration as a pre-agreed provision within a contract and thus be bound by the decision of the Arbitrator.

Arbitrators are normally appointed by one of THREE methods:

1   as a result of a pre-agreed decision between the Parties, normally through a provision within a contract to go to arbitration in the event of a dispute
    In this case, the Arbitrator itself may be named or subject to nomination by a professional body such as the Chartered Institute of Arbitrators.
2   by Statute, legislation in many countries includes the provision for disputes to be settled by arbitration
3   by Order of a Court.

The Arbitration Act is fairly brief and sets out procedures in the absence of any agreement to the contrary between the Parties.

As arbitration is a more flexible arrangement than legal proceedings, whatever procedure the Arbitrator and both Parties agree is sufficient in an arbitration case.

Whilst most forms of contract provide for resolution of disputes by arbitration, it must be stressed that it should only be seen as a last resort when all attempts at negotiation and other resolution methods have failed.

It may be surprising to hear that arbitration is common as a settlement procedure for disputes under trade agreements, maritime and insurance, consumer matters such as package holidays, construction industry disputes and property valuations.

The contracting Party that initiated that the dispute be referred to arbitration is always referred to as the "Claimant", whilst the other Party is referred to as the "Respondent".

It is important that these titles are clarified as the Arbitration Rules repeatedly refer to them.

Legal proceedings

It is not proposed to go into any detail about the process from initiation to judgement by a court, first as the contract does not give any details of the process, but also the proceedings themselves will depend on the law of the contract, which in turn is dependent on the location of the project.

However, the process will normally involve the Claimant's representative issuing a claim form or writ to the appropriate court; papers are then served on the Defendant.

The Defendant then replies to the service by either admitting or denying the claim against him; if he denies he replies by sending his defence which may also be accompanied by any counter-claim.

The case is then allocated to a court. There then follows a period of accumulation, collation and presentation of evidence, and hearings, again the process, the submission and the form of the hearing being dependent upon the law of the contract, until the court passes judgement.

Cases in most UK courts will normally have barristers acting as advocates on behalf of the Parties alongside the solicitors who have advised the Parties.

Chapter 11

Preparing and assessing tenders
Completing Scope, Site Information and Contract Data, inviting tenders, etc.

Question 11.1 What documents do we need to compile to invite tenders for an NEC4 Engineering and Construction Contract using Option B (priced contract with bill of quantities)?

Whilst any list of contract documents cannot be exhaustive as it depends on the type of project, the scope of works, requirements of the Client, etc., there are certain documents that will be required as a minimum when inviting tenders for an Engineering and Construction Contract using Option B.

These will normally consist of:

(i) Invitation to Tender (including instructions to tenderers on time and place of tender submission)

There is no "pro forma" version of "Instructions to Tenderers" within the NEC contracts, but any Invitation to Tender will give a brief description of the work, who it is for, and how tenders may be submitted.

Other inclusions will cover such matters as:

- time, date and place for the delivery of tenders
- what documents must be included with the tender, including a programme
- the policy regarding alternative and/or non-compliant bids
- arrangements for visiting the Site including contact details
- rules on non-compliant bids
- anti-collusion certificate.

(ii) Form of Tender

There is no "pro forma" Form of Tender within the NEC contracts, though the Engineering and Construction Contract Guidance Notes include a sample form. This is the tenderer's written offer to execute the work in accordance with the tender documents.

The Form of Tender is normally in the form of a letter with blank spaces for tenderers to insert their name and other particulars, total tender price, and other particulars of their offer. It is essential to have a standard Form of Tender and that all tenderers consistently use the same form, to make comparison of tenders easier.

(iii) The Pricing Document, in this case will be a Bill of Quantities prepared in accordance with the relevant Standard Method of Measurement as defined within Contract Data Part 1

(iv) Scope

Scope is defined by Clause 11.2(16) as

> *information which specifies and describes the works or states any constraints on how the Contractor Provides the Works and is either in the documents which the Contract Data states it is in, or in an instruction given in accordance with this contract.*

The Client provides the Scope and refers to it in Contract Data Part 1; if the Contractor provides Scope, e.g. for his design, this is included in Contract Data Part 2.

The main documents within the Scope are normally the drawings and specifications, but will also include:

Description of the works

- a statement describing the scope of the works
- schematic layouts, plan, elevation and section drawings, detailed working
- and/or production drawings (if relevant and available), etc.
- a statement of any constraints on how the Contractor Provides the Works, e.g. restrictions on access, sequencing or phasing of works, security issues, etc.

Plant and Materials

- materials and workmanship specifications
- requirements for delivery and storage
- future provision of spares, maintenance requirements, etc.

Health and safety

- specific health and safety requirements for the Site which the Contractor must comply with, particularly if the Site is within existing premises, including house and local safety rules, evacuation procedures, etc.
- any pre-construction information and health and safety plans for the project.

Financial Records

- details of any accounts and records to be kept by the Contractor.

Contractor's design

- the default position is that all design will be carried out by the Client.

If any or all of the design is to be carried out by the Contractor, it should be included within the Scope, together with any performance requirements to be met within the Contractor's design. In addition, any warranty requirements and also any future novation requirements should be included within the Scope.

Any design acceptance procedures should be included including time scales for submission and acceptance of design.

Completion

Completion is defined under Clause 11.2(2), but the Scope should also define any specific requirements that are required in order for completion to have taken place for example, and requirement for:

- "as built" drawings
- maintenance manuals
- training documentation
- test certificates
- statutory requirements and/or certification.

Services

- details of other Contractors and Others who will be occupying the Working Areas during the contract period and any sharing requirements.

Subcontracting

- lists of acceptable subcontractors for specific tasks
- statement of any work that should not be subcontracted
- statement of any work that is required to be subcontracted.

Programme

- any information that the Contractor is required to include in the programme in addition to the information shown in Clause 31.2. Also, if the Client requires the Contractor to produce a certain type of programme or to submit it using a certain brand of software, then this should be clearly detailed within the Scope.

Tests

- description of tests to be carried out by the Contractor, the Supervisor and Others, including those that must be done before Completion
- specification of materials, facilities and samples to be provided by the Contractor and the Client for tests
- specification of Plant and Materials that are to be inspected or tested before delivery to the Working Areas
- definition of tests of Plant and Materials outside the Working Areas that have to be passed before marking by the Supervisor.

Title

- statement of any materials arising from excavation or demolition to which the Contractor will not have title (Clause 73.2)
- requirements for Plant and Material to be marked (Clause 71.1).

Others

- There are also certain specific requirements for statements to be made in the Scope from certain Main and Secondary Options in the conditions of contract:
    - X4.1 the form of any ultimate holding company guarantee
    - X13.1 the form of any performance bond
    - X14.2 the form of any advance payment bond.

The Scope must be carefully drafted in order to define clearly what is expected of the Contractor in the performance of the contract and therefore included in the quoted tender amount and programme. If the contract does not cover all aspects of the work, either specifically or by implication, that aspect may be deemed to be excluded from the contract.

A comprehensive, all-embracing description should therefore be considered for the scope of work clause, which should be supplemented by specific detailed requirements.

If reliance is placed solely on a very detailed scope description, an item may be missed from this detailed description and be the subject of later contention.

Where items of equipment are to be fabricated or manufactured off Site by others, it is advisable that the contract sets out the corresponding obligations and liabilities of the respective Parties, particularly if these are to form an integral or key part of the completed works.

The Scope describes clear boundaries for the work to be undertaken by the Contractor. It may also outline the Client's objectives and explain why the work is being undertaken and how it is intended to be used. It says what is to be done (and maybe what is not included) in general terms, but not how to do it or the standards to be achieved. It explains the limits, where the work is to interface with other existing or proposed facilities. It may draw attention to any work or materials to be provided by the Client or others. It should also emphasise any unusual features of the work or contract, which tenderers might otherwise overlook.

This is the document that a tenderer can look to, to gain a broad understanding of the scale and complexity of the job and be able to judge his capacity to undertake it. It is written specifically for each contract. In some respects it is analogous to a shopping list. It should be comprehensive, but it should be made clear that it is not intended to include all the detail, which is contained in the drawings, specifications and schedules.

The Scope will also include drawings which again should provide clear details of what the Contractor has to do. Clearly, tenderers must be given sufficient information to enable them to understand what is required and thus submit considered and accurate tenders.

It will also include the specification which is a written technical description of the standards and various criteria required for the work, and should complement the drawings. The specification describes the character and quality of materials and workmanship for work to be executed.

Again, it may lay down the order in which various portions of the work are to be executed. As far as possible, it should describe the outcomes required, rather than how to achieve them. It is customary to divide the work into discrete sections or trades (e.g. drainage, concrete, pavements, fencing, etc.), with clauses written to cover the materials to be used, the packaging, handling and storage of materials (only if necessary), the method of work to be used (only if necessary), installation criteria, the standards or tests to be satisfied, any specific requirements for completion, etc.

The specification is an integral part of the design. This is often overlooked, with the result that inappropriate or outmoded specifications are selected, or replaced by a few brief notes on the Drawings. The designer should spend an appropriate amount of time specifying the quality of the work, as it is not possible to price, build, test or measure the work correctly unless this is done.

The Scope describes what the Contractor has to do in terms of scope and standards, and in some cases must not do or include in order to Provide the Works. It may also include the order or sequence in which the works are to be carried out. It also details where the work is to interface with other existing or proposed Contractors or facilities. It will also include any work or facilities or materials to be provided by the Client or others. It may also include any unusual features of the work, which tenderers might otherwise overlook, for example any planning constraints, etc.

This is the document from which a tenderer can gain a broad understanding of the scale and complexity of the job and his capacity to undertake it. It is written specifically for each contract.

The Scope should be:

- clear – unambiguous
- concise – not excessively wordy
- complete – have nothing missing.

In respect of the Contractor's design, a reason for not accepting the design is that it does not comply with either the Scope or the applicable law. Again, if the Scope was lacking and the Contractor provided a design that complied with it, and the applicable law, could the Contractor escape liability for the defective design?

Secondary Option X15 states that "the Contractor is not liable for a Defect which arose for its design unless it failed to carry out that design using the skill and care normally used by professionals designing works similar to the works".

(v) Site Information

Site Information is defined in Clause 11.2(18) as *"information which describes the Site and its surroundings and is in the documents which the Contractor states it is in"*, and is identified in Contract Data Part 1.

Site Information may include:

1. Ground investigations, borehole and trial pit records and test results. If the Client has obtained such information, it should not be withheld from the tenderers, though the tenderers should be aware that the Site Information alone cannot be relied upon in terms of a possible compensation event (see Clause 60.2).
2. Information about existing buildings, structures and plant on or adjacent to the Site.
3. Details of any previously demolished structures and the likelihood of any residual surface and subsurface materials.
4. Reports obtained by the Client concerning the physical conditions within the Site or its surroundings. This may include mapping, hydrographical and hydrological information.
5. All available information on the topography of the Site should be made accessible to tenderers, preferably by being shown on the Drawings.
6. Environmental issues, for example nesting birds and protected species.
7. References to publicly available information.
8. Information from utilities companies and historic records regarding plant, pipes, cables and other services below the surface of the Site.

It is vital that care is taken to get the Site Information correct. Tenderers must be given sufficient information to enable them to understand what is required and thus to submit considered and well-priced tenders.

Under Clause 60.3, if there is an ambiguity or inconsistency within the Site Information, the Contractor is assumed to have taken into account the conditions most favourable to doing the work. Whilst many interpret that as the Contractor allowing the cheapest way of doing the work, it may be the easiest or quickest way.

In the event of the Contractor notifying a compensation event under Clause 60.1(12) for unforeseen physical conditions, he is assumed to have taken into account the Site Information, publicly available information referred to in the Site Information, information obtained from a visual inspection of the Site, and other information that an experienced contractor could reasonably be expected to have or to obtain. So the Contractor cannot rely solely on the Site Information in terms of how he prices and programmes the works.

The Contract Data

The Contract Data provides the information required by the conditions of contract specific to a particular contract. Other contracts call this the "Appendices" or "Contract Particulars".

Contract Data Part 1

Part 1 consists of data provided by the Client, the sections of the Contract Data aligning with the sections of the core clauses.

The Contract Data requires the Client, or the Party representing the Client, to identify:

Section 1 – General

- the selected Main and Secondary Options
- a description or title for the works
- the names and contact details of the Client, the Project Manager, the Supervisor and the Adjudicator.

Whilst the Project Manager and the Supervisor must always be named individuals, many clients choose to insert company names for these Parties within the Contract Data and to separately identify the named individuals. It is also not uncommon for clients to name directors or partners of their respective companies, then the authority of the Project Manager and Supervisor is delegated to the individuals under Clause 14.2.

- The Scope and Site Information are also identified within this section, though these are normally incorporated by reference to separate documents rather than listing drawing numbers and specification references

within the Contract Data. If this is the case, the separate documents must be clearly defined.
- The boundaries of the site are defined, normally by reference to a specific drawing or map.
- As the NEC4 contracts are intended for use worldwide, the language and the law of the contract are also entered.
- The period for reply is the period that the parties have to reply to submissions, proposals, notifications, etc. where there is no period specifically stated elsewhere within the contract; for example, the period within which a party should reply to an early warning notice, or the Project Manager should reply to the contractor submitting the particulars of its design or the name of a proposed Subcontractor.

Different periods for reply can be inserted into Contract Data Part 1 for different communications; also they can be mutually extended by agreement between the communicating parties.

The period of reply would clearly not apply to the Contractor's submission of a programme for acceptance or the submission of a quotation for a compensation event.

- There is reference to matters to be included within the Early Warning Register and the frequency of early warning meetings.

Section 2 – Contractor's main responsibilities

- Key Dates and Conditions are inserted (if applicable)
- The frequency of the requirement for the Contractor to prepare forecasts of the total Defined Cost for the whole of the works (if Option C, D, E or F is used).

Section 3 – Time

- The starting date
- Access date(s)
- The frequency of requirement for the Contractor to submit revised programmes are inserted within this section. Many clients change the reference to "one calendar month" rather than "weeks", to align with monthly progress meetings or reporting requirements.
- The completion date for the whole of the works. If the Client has decided the completion date for the whole of the works, the date is inserted here, alternatively the tendering contractors may be required to insert a date in Contract Data Part 2.
- Whether the Client is prepared to take over the works before the Completion Date.

When the Contractor completes the works, the Project Manager certifies Completion and the Client takes over the works not later than two weeks after Completion, even though takeover may be before the Completion Date. If the Client is not willing to take over the works before the Completion Date, the statement to that effect should remain in the Contract Data, if not it should be deleted. This statement is referred to in Clause 35.1.
- The period after the Contract Date when the Contractor is to submit a first programme for acceptance.

Section 4 – Testing and Defects

- The period after the Contract Date when the Contractor is to submit a quality policy statement and a quality plan.
- The defects date is identified as a number of weeks after Completion of the whole of the works. This identifies the period the contractor is initially liable for correcting defects.
- The defect correction period is also stated, this being the period within which the contractor must correct each notified defect, failing which the Project Manager assesses the cost to the client of having the Defect corrected by Others.

The Contract Data provides for three entries to be inserted here if required. A defect correction period for the whole of the works, with two further entries for alternative periods for specific parts of the works, dependent on the urgency to have them corrected, for example:

All works except for ...	7 days
Security installations	24 hours
Other mechanical and electrical installations	48 hours

Clearly, any defect that has a potential effect on health and safety should be corrected as soon as possible.

Section 5 – Payment

- Again, as the NEC4 contracts are intended for use worldwide, the currency of the contract is entered.
- The assessment interval is also entered, which again is often expressed as "one calendar month" rather than in "weeks". The Project Manager decides

the first assessment date to suit the Parties and following assessments are carried out within the assessment intervals.
- The interest rate on late payments is stated.
- The period in which payments are made (if not three weeks).
- The share percentages (if Option C or D is used).
- The exchange rates (if Option C, D, E or F is used).

Section 6 – Compensation events

- There are entries within this section in respect of the weather, the place where weather is to be recorded (weather station, airport, etc.), the weather measurements to be recorded, the default measurements being cumulative monthly rainfall, the number of days with rainfall more than 5mm, minimum air temperature less than 0 degrees Celsius, and snow lying at the designated time of day. The supplier of the weather measurements, the place where weather measurements are recorded, and where they are available from are also stated.

  For some isolated site locations where no recorded data may be available, assumed values may be inserted.
- The value engineering percentage is 50% by default unless another percentage is stated.
- The method of measurement (if Option B or D is used).
- Additional compensation events.

Section 7 – Title

This section is not included within the Contract Data as no entries are required.

Section 8 – Risks and insurance

- Any additional Client's liabilities are stated.
- The minimum limit of indemnity for third party public liability and the Contractor's liability for its own employees is stated within this section.

Contract Data Part 1 then contains a series of optional statements which are completed where appropriate:

- If the Client is to provide Plant and Materials, there is provision for insurance of the works to include any loss or damage to such Plant and Materials.
- The Contractor provides the insurances stated within the Insurance Table, except any insurances which the Client is to provide which are stated in Contract Data Part 1, which lists what the insurance is to cover, the amount of cover and the deductibles (otherwise known as "excesses"), the amounts to be paid by the insuring Party, often before the insurance company pays.

218   Preparing and assessing tenders

Contract Data Part 1 then contains a series of entries related to resolving and avoiding disputes:

- The tribunal is stated.
- If the tribunal has been identified as arbitration, the Client must identify the arbitration procedure, the place where any arbitration is to be held, who will choose an arbitrator if either of the Parties cannot agree a choice, or if the arbitration procedure does not state who selects an arbitrator.
- The Senior Representatives of the Client are stated.
- The name and contact details of the Adjudicator are stated.
- With regard to dispute resolution, the Adjudicator nominating body is named within this section, normally a professional institution, and then the tribunal is named as either arbitration or legal proceedings.

If Option W3 is used:

- the number of members of the Dispute Avoidance Board is stated (either one or three members)
- the Client's nomination for the Dispute Avoidance Board is stated, also the frequency of visits to Site by the Dispute Avoidance Board, and the Dispute Avoidance Board nominating body
- finally, dependent on which secondary options are selected, the Client enters details for the relevant entry in the Contract Data.

Contract Data Part 2

Part 2 is completed by the Contractor and includes:

- the name and contact details of the Contractor
- the fee percentage
- names and details of key people
- any matters to be included in the Early Warning Register
- any Scope provided by the Contractor for its design
- details of any programme submission
- completion date if decided by the tenderer
- details of the price and the relevant pricing document, i.e. activity schedule or bill of quantities
- names of Senior Representatives
- any entries in respect of Secondary Options
- rates and percentages in respect of the Schedules of Cost Components.

Question 11.2   We are preparing the Bill of Quantities for a project to be carried out under Option B (priced contract with bill of quantities). Is the preparation of the Bill of Quantities any different to any

other contract? The Client wishes to make a decision about certain landscaping elements as the project proceeds, so we are intending to include Provisional Sums within the Bill of Quantities for the time being.

First, the preparation of a Bill of Quantities for an NEC4 Engineering and Construction Contract Option B or Option D is the same as for any other contract, so the same Method of Measurement will be used as defined within Contract Data Part 1 and the quantities would be entered into the Bill of Quantities in the normal way.

However, one would have to give some thought as to the normal "Preambles" and "Preliminaries" sections as they would need to be labelled as "Scope" (information that "specifies and describes the works or states any constraints on how the Contractor Provides the Works") and "Site Information" ("information which describes the Site and its surroundings").

With regard to the question about using Provisional Sums for certain landscaping elements, let us first consider the definition of Provisional Sums.

Provisional Sums are used where there are elements of work which are not designed or cannot be sufficiently defined at the time of tender and therefore a sum of money is included by the Client in the Bill of Quantities or other pricing document to cover the item.

When the item is defined or able to be properly defined the Contractor is given the information which allows him to price it, the Provisional Sum is omitted and the price included in its stead.

The problem with Provisional Sums is that they reduce the competition amongst tenderers as they are not priced at tender stage, and also if they are "defined" the Contractor is deemed to have allowed time in his programme for them, if they are "undefined" he has not.

The NEC contracts do not, and have never provided, for Provisional Sums, the principle being that:

1 the Client decides what he wants at tender stage so it can be designed and accurately described and the tendering Contractors can properly price and programme for them; or
2 when the Client decides what he wants, the Project Manager can give an instruction to the Contractor which changes the Scope; it is a compensation event under Clause 60.1(1) and can be priced at the time. In that case, the cost of the work is held in the *scheme* rather than in the *contract*.

Index

Accepted Programme x–xi, 4, 12, 38, 105–24, 162–63, 165, 169–71, 174, 189; first x, 110–11, 113
Adjudicator xvi 15–16, 24–25, 31, 115, 149, 167–68, 174, 194–205, 214, 218; named 201; new 195, 198; nominating 202; qualified 203
Adjudicator Nominating Bodies 202–3, 218
Adjudicator's decision 196, 199–200
Adjudicator's fees 25
Advanced payments 34, 135
Affected Property 20–21, 83, 85, 178, 186
Alliance Contract (ALC) 1–2, 64–66, 86
Arbitrator 196, 199, 202, 205–6, 218

Base Date Index 60
Building Cost Information Service 77
Building Information Modelling (BIM) 3

Centre for Effective Dispute Resolution (CEDR) 202
Chartered Institute of Arbitrators (CIArb) 202, 206
Client's Agent 24, 41
Client's Designers 49, 91
Client's liabilities, main categories of 182, 186
Completion x, 32–33, 38–39, 47, 68–72, 105–10, 123, 126–30, 143–44, 146–47, 163–66, 169–70, 173, 210–12, 216; earliest 111, 116; early 87, 106; late 84
Completion Certificate 129
Completion Dates 45, 105–8, 111–14, 116, 122–24, 132, 159–60, 164, 166, 168–70, 174, 178, 181, 215–16, 218; current 112, 116; earlier 68, 124, 132, 169; planned xi, 59, 112, 114, 116, 126, 128, 143–44, 216

Completion Dates and Key Dates 124, 178
Constructing the Team 52
Construction Management Option 41
Contract Data 4 7–8, 24–25, 33–36, 38–39, 54–56, 68–71, 75–79, 91–93, 102, 106–7, 109–11, 113–14, 134–36, 145–47, 149, 153–154, 159–160, 185–186, 199–202, 204–5, 209, 213–15, 217–19
Contract Date 25, 33, 39, 61, 93, 110–11, 113, 133, 139–40, 146, 157, 162–64, 171, 191, 216
Contractor forecasts of total Defined Cost 47–48
Contractor's accounts and records 47–48, 141, 143–44
Contractor's design viii–ix, 3–4, 13, 15, 18, 42, 44, 53–54, 56, 67, 99–104, 125–26, 131, 210, 213
Contractor's design of Equipment 44
Contractor's insurance policies and certificates 46
Contractor's liability for correction of latent defects 130
Contractor's plan 21, 84
Contractor's programme 5, 18, 45, 56, 111–12, 115, 120
Contractor's Proposals ix, 5, 44, 48, 101, 178
Contractor's quality policy statement and quality plan 45
Contractor's quotation 45–46, 124, 174, 179
Contractor's risk 59, 111, 115, 117, 160
Contractor's Scope 54, 102
Contractor's share xiii, 22, 47, 70–72, 127, 146–47, 150
Contractor's share percentages 147

Index

Contractor's submission 84, 110, 113, 120, 174, 215
Contractor's time risk allowances 117
Core Group 63–64
Corrupt Acts 5, 190

damage to Plant and Materials and equipment 186
DBFO (design-build-finance-operate) 2, 28
DBO *see* Design Build and Operate
Defects Certificate xi, 7, 36, 50–51, 56, 70, 128–30, 134, 146, 150, 166–67, 173, 184
Defined Cost 22, 39–40, 59–61, 64–65, 72–73, 79–80, 126, 134–36, 139–42, 146, 148–49, 152–54, 168–70, 173, 175; actual 6, 156, 168–69; forecast 156, 168–69, 175, 190–91
Defined Cost for compensation events 61
Defined Cost for Plant and Materials 191
Defined Cost of subcontracted work 136
delay damages 14, 37
delayed Completion 121
Design and Build 27
Design Build and Operate (DBO) vi, 1–2, 27–28, 86
Disallowed Costs xiii, 22, 40, 47–48, 80, 126, 142–46
Dispute Avoidance Board 8, 144–45, 168, 193, 200–1, 203–4, 218
Dispute Reference Table 194, 198
Dispute Resolution Board 8
dispute resolution Option W1 193
dispute resolution Option W2 196
dispute resolution Option W3 200
Dispute Resolution Service Contract (DRSC) 1, 195, 198, 203
Dispute Resolver 204–5

Early Contractor Involvement *see* ECI
Early Warning Register ix, 9, 91–98, 215, 218; first 44, 96; revised 97
ECC *see* Engineering and Construction Contract
ECC acceleration 123
ECI (Early Contractor Involvement) 4, 38, 62
ECS (Engineering and Construction Subcontract) 1, 86
ECSC *see* Engineering and Construction Short Contract

ECSS (Engineering and Construction Short Subcontract) 1
Egan 62
Employer's Liability Insurance 184–86
Engineering and Construction Contract (ECC) vi–xvii, 3–13, 16–21, 38–44, 55–57, 80–82, 99–102, 105–13, 128–31, 134–39, 157–62, 166–68, 170–74, 176–80, 187–91
Engineering and Construction Short Contract (ECSC) ix, 1, 9–10, 17–19, 32, 81–82, 99, 158, 161

FC (Framework Contract) 1
fee percentage 39, 146
FIDIC contracts 10
Financial Records 210
FM (facilities management) 2, 21
Framework Contract (FC) 1

Goods Information 9

Housing Grants, Construction and Regeneration Act 1996 vi, 8, 14–15, 23, 29–31, 38, 42, 66, 87, 149, 192–93, 196–98, 200, 204

Incentive Schedule 70
Information Model 3
Information Plan 3
Information Providers 3
Information Required 115
Insurance Table 4, 183–87, 217
Interest on late payment 191

Key Dates 44–46, 90, 109–10, 112, 114–16, 123–24, 151, 161, 164, 166, 168–70, 178
Key Performance Indicators *see* KPIs
KPIs (Key Performance Indicators) viii, 38, 48, 50, 63–64, 70

Latest Index 60
Latham 62
Latham Report Constructing the Team 17
Local Democracy, Economic Development and Construction Act 2009 vi, 15, 30–31, 42, 149, 192–93, 197

Main Options vi, 14, 16, 18, 22–23, 29–30, 37, 39, 41, 81, 148, 150–51, 153, 176, 188

Index

Management Contract vi, 10, 14, 19, 41, 80–82, 153
marking Plant and materials 49–50

Named Suppliers 133–34
naming Subcontractors vii, 56
naming specialist Subcontractors 56
Naming Subcontractors 57
NEC3 Adjudicator's Contract 8, 203
NEC3 contracts 4, 6–7, 9, 38, 63, 67, 75, 93, 96, 122, 144–45, 162–63, 165, 178
NEC3 Engineering and Construction Contract vi, 2–3, 6, 35, 80, 136, 200
NEC3 Professional Services Contract 2, 6, 35, 153
NEC4 Alliance Contract 1–2, 64–65
NEC4 Contracts vi–xvi, 1–86, 88, 101, 103, 105–6, 108–9, 130, 133, 162, 176–77, 179, 182, 205, 215–16
NEC4 Design Build and Operate Contract 1–2, 28
NEC4 Dispute Resolution Service Contract xvi, 1, 8, 195, 198, 200, 203
NEC4 Engineering and Construction Contract vi–xvii, 4–7, 9–10, 38–42, 80–82, 99–102, 105–6, 108–11, 120–23, 128, 134–36, 149–51, 157–59, 169, 172–74, 178–80
NEC4 Engineering and Construction Contract programme 21
NEC4 Engineering and Construction Short Contract ix, 1, 9, 17–18, 82, 99, 158
NEC4 Engineering and Construction Short Subcontract 1
NEC4 family 1–2, 32, 42, 64
NEC4 Framework Contract 1
NEC4 Professional Service Contract xiv–xv, 1, 6, 36–38, 41, 153–54, 158, 171, 182, 187
NEC4 Professional Service Short Contract 1, 36, 38
NEC4 Professional Service Subcontract 1
NEC4 Short Contracts 27, 81
NEC4 Supply Contract 1, 6, 25, 27
NEC4 Supply Short Contract 1, 25–27
NEC4 Term Service Contract ix, 1, 6, 20–21, 24, 81, 83, 159, 182
NEC4 Term Service Short Contract 1, 20, 24
NEC Adjudicator's Contract 203
NEC family 1, 28, 203
NEC Panel 133
New Secondary Options 2, 178
Nominating Subcontractors 57
notified Defects 129, 131, 137, 216; corrected, 106, 122, 108

OGC (Office of Government Commerce) 133
Option W1 xvi, 14, 22, 29, 37, 167, 193, 198, 203
Option W2 14, 22, 29–30, 37, 167, 193, 196, 198
Option W3 8, 168, 193, 200, 218
Option X1 Price adjustment for inflation 14, 23, 26, 37, 39–40, 59–61, 146
Option X2 Changes in the law 23, 26, 61, 66
Option X3 Multiple currencies 14, 23, 26, 29, 37, 86, 135
Option X4 Ultimate holding company guarantee 14, 23, 26, 29, 37, 59, 66, 86
Option X5 Sectional completion 86, 106
Option X6 Bonus for early completion 87, 106
Option X7 Delay damages 26, 70, 87, 106–8, 114
Option X8 Undertakings to the Client or Others 2, 14, 23, 29, 35, 37
Option X9 Transfer of rights 14, 29, 38, 66, 87
Option X10 Information Modelling 3, 14, 23, 26, 29, 38, 66, 87
Option X11 Termination by the Client 14, 23, 26, 38, 87
Option X12 Multiparty collaboration 26, 63–65, 87
Option X13 Performance bond 15, 23, 26, 29, 38, 59, 87
Option X14 Advanced payment to the Contractor 15, 26, 29, 87
Option X15 The Contractor's design 3, 67, 87
Option X16 Retention 15, 36, 87, 130, 134
Option X17 Low performance damages viii, 15, 68, 70, 87
Option X17 Low service damages 23
Option X18 Limitation of liability 15, 23, 26, 29, 66, 87
Option X19 Termination 23, 87
Option X20 Key Performance Indicators 26, 70, 87

Index

Option X21 Whole life cost 15, 26, 87, 178
Option X22 Early Contractor involvement 15, 38, 66, 87
Option X23 Extending the service period 23, 29
Option X24 The accounting periods 87
Option X25 Supplier warranties 26, 87
Option X26 Programme of work 66, 87
Option Z Additional conditions 87

PAF 60–61
Partnering 9, 62–63; defined 62
Partnering Information 63–64
Partners 2, 63–66, 214
People Rates 7, 74, 77, 79, 155
PFI contracts 31
PI (Professional Indemnity) xvi, 68, 184–87
planned Completion 111–17, 169–71
Plant and Materials 19, 21, 46, 50–51, 76, 78–79, 117–18, 144, 148, 183, 186, 191, 209, 211, 217; cost of disposal of 76, 78; fabrication of 77–78
PPP (Public Private Partnership) 27
Preliminaries 138, 219
Price Adjustment Factor 60–61
Price adjustment for inflation 23, 26
Price for Work 18, 36, 60–61, 72–73, 83, 120, 134, 136–38, 142, 147–48, 150–53, 190–91
Price for Work Done to Date (PWDD) 18, 36, 60–61, 71–73, 83, 120, 134, 136–38, 142, 147–48, 150–53, 190–91
price list Option 22, 81
Price Lists ix, 18–19, 22–24, 27–28, 81–84
Prime Cost Sums 57
Professional Indemnity *see* PI
Professional Indemnity claims 187
Professional Service Contract (PSC) 1, 6, 37, 86, 158, 161, 171, 185
Professional Service Short Contract (PSSC) 1, 36, 38
Professional Service Subcontract (PSS) 1, 86
Project Bank Account xii, 15, 23, 26, 38, 66, 133–34
Project Information 3
Project Manager xi–xv, 3–8, 41–44, 49–56, 88–94, 100–2, 104–7, 118–29, 131–32, 134–37, 140–53, 164–67, 172–75, 177–82, 194–96

Project Manager's acceptance x, 54, 102, 104
Project Manager's role 42
Proof of Insurance 185
Provisional Sums xv, xvii, 172, 219
PSC *see* Professional Service Contract
PSS (Professional Service Subcontract) 1, 86
PSSC (Professional Service Short Contract) 1, 36, 38
Public Liability Insurance 185–86
Purchaser's Acceptance 27
PWDD *see* Price for Work Done to Date

Quality management xi, 6, 13, 21, 26–27, 29, 37, 66, 125–32
Quantity Surveyors 43–44

Rights of Third Parties 3, 15, 23, 26, 29, 38, 66, 87
Risk Register in NEC3 93
Risk Registers ix, 9, 44, 88, 90–93, 95–98

SC (Supply Contract) 1, 6, 27, 86
SCC *see* Schedule of Cost Components
Schedule of Basic Plant Charges 77
Schedule of Core Group Members 63–64
Schedules of Cost Components viii–ix, 6, 9, 18, 30, 55, 59, 66, 72–74, 76–80, 124, 134–36, 154–55, 164, 218
Schedule of Partners 63–64
Scheme for Construction Contracts 67, 197, 204
Scope xvii, 2, 3, 5, 9, 18–20, 24, 27, 33, 35, 36, 38, 39, 44–45, 48–50, 53, 54, 56, 57–58, 67–69, 84–85, 90, 99–102, 104, 106–109, 112–13, 115, 119, 125, 128, 131–32, 134, 139, 141, 143–45, 149, 159, 165, 174–78, 182, 208–14, 218–19
search for a Defect xi, 128
Secondary Options 1, 4, 12, 14, 17–18, 23, 37, 42, 63, 66–67, 81, 86, 211, 214, 218
Sectional Completion x, 106, 109
Senior Representatives 66, 167, 194, 198, 218; nominated 8
Service Information 9
Service Manager 21, 52, 65, 83–84, 156, 187
Shorter Schedule of Cost Components (NEC3) 80
Short Schedule of Cost Components (NEC4) viii, 6, 55, 59, 72–73, 77, 79–80, 154–55

The Site and Working Areas 21
Site Information 58, 158, 213–14, 219
SSC (Supply Short Contract) 1, 27
Subcontractor costs 6, 80
Subcontractors 3, 6, 18, 35, 56–57, 76, 78–80, 100, 115, 117–18, 141, 143, 145, 155, 189–90; payments to 76, 78–79, 155; procurement of 143, 145
Subcontractor's payment 76
Submitting design proposals ix, 99
Suggested template for early warning 89
Supervisor vii, 18, 21, 24, 43–44, 48–53, 56, 72, 91, 100, 126–30, 164, 169, 211, 214
Supply Contract see SC
Supply Manager 25
Supply Short Contract (SSC) 1, 27

Take over 106
Task Completion 83
Task Completion Date 83–85
Task Order programme 83–85
Task Orders ix, 83–85
Task price list 84
Task Schedule 154
Task Starting Date 84

tendering Contractors xiii, 34, 41, 58, 100, 102, 138, 201, 215, 219
Termination Table 190
Term Service Contract (TSC) ix, 1, 6, 20–21, 23–24, 81, 83, 86, 159, 162, 182, 186–87, 195, 199
Term Service Contracts 20–21, 23, 86, 159, 162, 186–87
Term Service Short Contract (TSSC) 1, 24
Time Charge 153–54
Time Charge and expenses 153
Title to Plant and Materials xvi, 180
Total Contractor's share 22, 147
total Defined Cost 4, 38, 47–48, 72, 80, 152–53, 156, 177, 215
TSC see Term Service Contract
TSSC see Term Service Short Contract
Typical Early Warning Register 94

VAT 83

Withholding acceptance 56
Working Areas vii, 7, 20–21, 41, 49, 51, 54–55, 73–74, 76–78, 136, 144, 148, 191, 210–11; extended 55
Works Contractors 2, 41, 64
Works Information 9, 177